Debbie

THE 100+ SERIES™

Common Core Edition

PRE-ALGEBRA

Essential Practice for Advanced Math Topics

D1371619

Carson-Dellosa Publishing, LLC
Greensboro, North Carolina

Visit carsondellosa.com for correlations to Common Core, state, national, and Canadian provincial standards.

Carson-Dellosa Publishing, LLC
PO Box 35665
Greensboro, NC 27425 USA
carsondellosa.com

© 2014, Carson-Dellosa Publishing, LLC. The purchase of this material entitles the buyer to reproduce worksheets and activities for classroom use only—not for commercial resale. Reproduction of these materials for an entire school or district is prohibited. No part of this book may be reproduced (except as noted above), stored in a retrieval system, or transmitted in any form or by any means (mechanically, electronically, recording, etc.) without the prior written consent of Carson-Dellosa Publishing, LLC.

Printed in the USA • All rights reserved.

ISBN 978-1-4838-0076-9

04-137167784

Table of Contents

Introduction

What are the Common Core State Standards for Middle School Mathematics?

In grades 6–8, the standards are a shared set of expectations for the development of mathematical understanding in the areas of ratios and proportional relationships, the number system, expressions and equations, functions, geometry, and statistics and probability. These rigorous standards encourage students to justify their thinking. They reflect the knowledge that is necessary for success in college and beyond.

Students who master the Common Core standards in mathematics as they advance in school will exhibit the following capabilities:

1. Make sense of problems and persevere in solving them.

Proficient students can explain the meaning of a problem and try different strategies to find a solution. Students check their answers and ask, "Does this make sense?"

2. Reason abstractly and quantitatively.

Proficient students are able to move back and forth smoothly between working with abstract symbols and thinking about real-world quantities that symbols represent.

3. Construct viable arguments and critique the reasoning of others.

Proficient students analyze problems by breaking them into stages and deciding whether each step is logical. They justify solutions using examples and solid arguments.

4. Model with mathematics.

Proficient students use diagrams, graphs, and formulas to model complex, real-world problems. They consider whether their results make sense and adjust their models as needed.

5. Use appropriate tools strategically.

Proficient students use tools such as models, protractors, and calculators appropriately. They use technological resources such as Web sites, software, and graphing calculators to explore and deepen their understanding of concepts.

© Carson-Dellosa • CD-704384

6. Attend to precision.

Proficient students demonstrate clear and logical thinking. They choose appropriate units of measurement, use symbols correctly, and label graphs carefully. They calculate with accuracy and efficiency.

7. Look for and make use of structure.

Proficient students look closely to find patterns and structures. They can also step back to get the big picture. They think about complicated problems as single objects or break them into parts.

8. Look for and express regularity in repeated reasoning.

Proficient students notice when calculations are repeated and look for alternate methods and shortcuts. They maintain oversight of the process while attending to the details. They continually evaluate their results.

How to Use This Book

In this book, you will find a collection of 100+ reproducible practice pages to help students review, reinforce, and enhance Common Core mathematics skills. Use the chart provided on the next page to identify practice activities that meet the standards for learners at different levels of proficiency in your classroom.

© Copyright 2010. National Governors Association Center for Best Practices and Council of Chief State School Officers. All rights reserved.

Common Core State Standards* Alignment: Pre-Algebra

Domain: Ratios and Proportional Relationships	
Standard	**Aligned Practice Pages**
6.RP.A.1	7, 8
6.RP.A.2	10, 11
6.RP.A.3b	10, 11
6.RP.A.3c	12–15
7.RP.A.1	10
7.RP.A.2a	9
7.RP.A.2b	10, 11
7.RP.A.2c	9, 11

Domain: The Number System	
Standard	**Aligned Practice Pages**
6.NS.A.1	17–28
6.NS.B.3	31–40, 42, 44
6.NS.C.5	58
6.NS.C.6a	45
6.NS.C.6b	68–70
6.NS.C.6c	68–70
6.NS.C.7a	65, 84
6.NS.C.7b	65, 84
6.NS.C.7c	66
6.NS.C.7d	66
6.NS.C.8	68–70
7.NS.A.1b	45–48, 51
7.NS.A.1c	49–51, 66
7.NS.A.1d	45–51, 57–60, 67
7.NS.A.2a	52, 53, 61–63, 67
7.NS.A.2b	54–56, 61–63
7.NS.A.2c	52–58, 61–63, 67
7.NS.A.2d	29, 30, 41, 43
7.NS.A.3	46–63, 65–67

Domain: Expressions and Equations	
Standard	**Aligned Practice Pages**
6.EE.A.1	64, 72
6.EE.A.2a	82, 83, 102
6.EE.A.2c	75–77
6.EE.A.3	73, 74, 78–81
6.EE.A.4	73, 74, 78–81
6.EE.B.5	76
6.EE.B.6	82, 83, 102
6.EE.B.7	83
6.EE.B.8	84
6.EE.C.9	103
7.EE.A.1	73, 74, 78–80
7.EE.A.2	73, 74, 78–80, 82
7.EE.B.3	75, 81, 83, 92–103
7.EE.B.4a	82, 102, 103
7.EE.B.4b	85–91

* © Copyright 2010. National Governors Association Center for Best Practices and Council of Chief State School Officers. All rights reserved.

Name________________________________

6.RP.A.1

Ratios

Write each ratio as a fraction in simplest form.

3 to 12 → $\frac{3}{12} = \frac{1}{4}$ 65 : 35 → $\frac{65}{35} = \frac{13}{7}$

6 out of 40 → $\frac{6}{40} = \frac{3}{20}$

1. 196 to 7

2. 19 : 76

3. 18 out of 27

4. $\frac{3}{8}$ to $\frac{3}{4}$

5. 0.11 : 1.21

6. 140 : 112

7. 18 to 27

8. 54 out of 87

9. 112 : 140

10. 88 to 104

11. 65 out of 105

12. 65 : 117

13. 165 to 200

14. 168 : 264

Name______________________________ 6.RP.A.1

More Ratios

I. Express each ratio as a fraction in simplest form.

45 seconds to 3 minutes.

$$\frac{45}{3 \cdot 60} = \frac{45 \text{ sec}}{180 \text{ sec}} = \frac{1}{4}$$

The ratio of males to total students in a school with 1200 males and 1000 females.

$$\frac{1200}{1000 + 1200} = \frac{1200 \text{ males}}{2200 \text{ students}} = \frac{6}{11}$$

1. 55¢ to $4
2. 10 inches to 1 yard
3. 8 hours to 3 days
4. The ratio of wins to losses in 35 games with 21 losses and no ties
5. The ratio of the area of a rectangle with sides of 6m and 8m to the area of a square with sides of length 12m
6. The ratio of girls to boys in a class of 40 students with 17 girls
7. Big Bob's batting average if he had 3 hits in 4 at bats against the Cougars
8. The ratio of wins to losses in 42 games with 35 wins and no ties

II. Express each ratio as a fraction in simplest form.

Find two numbers in the ratio of 4:3 whose sum is 63.

Let x = common factor	$4x + 3x = 63$	$4 \cdot 9 = 36$
$4x$ = first number	$7x = 63$	$3 \cdot 9 = 27$
$3x$ = second number	$x = 9$	The numbers are 36 and 27.

9. A 36cm segment is divided into three parts whose lengths have the ratio of 2:3:7. Find the length of each segment.
10. The sum of the measures of two complementary angles is 90°. Find the measures of two complementary angles whose measures are in the ratio of 1:4.

© Carson-Dellosa • CD-704384

Name________________________________

7.RP.A.2a, 7.RP.A.2c

Proportions

Solve each proportion.

$\frac{3}{7} = \frac{x}{49}$ $3 \cdot 49 = 7x$

$\frac{147}{7} = \frac{7x}{7}$ $21 = x$

1. $\frac{8}{6} = \frac{m}{27}$

2. $\frac{z}{3} = \frac{8}{15}$

3. $\frac{16}{40} = \frac{24}{c}$

4. $\frac{9}{p} = \frac{5}{2}$

5. $\frac{1.8}{x} = \frac{3.6}{2.4}$

6. $\frac{4}{5} = \frac{0.8}{y}$

7. $\frac{x}{2} = \frac{15}{5}$

8. $\frac{18}{12} = \frac{24}{x}$

9. $\frac{18}{15} = \frac{6}{x}$

10. $\frac{121}{x} = \frac{220}{100}$

11. $\frac{1.6}{x} = \frac{14}{21}$

12. $\frac{x}{168} = \frac{66\frac{2}{3}}{100}$

13. $\frac{x}{32} = \frac{37\frac{1}{2}}{100}$

14. $\frac{16}{48} = \frac{x}{100}$

15. $\frac{0.12}{.25} = \frac{x}{100}$

16. $\frac{1.5}{x} = \frac{0.07}{0.14}$

Name________________________________ 6.RP.A.2, 6.RP.A.3b, 7.RP.A.1, 7.RP.A.2b

Problems Using Proportions

Three loaves of bread cost \$3.87. How much do 2 loaves cost?

$\frac{\text{number of loaves}}{\text{cost}}$

$$\frac{3}{3.87} = \frac{2}{x}$$

$$3x = 2 \cdot 3.87$$

$$\frac{3x}{3} = \frac{7.74}{3}$$

$$x = 2.58$$

2 loaves cost \$2.58

1. If 64 feet of rope weigh 20 pounds, how much will 80 feet of the same type of rope weigh?

2. If a 10 pound turkey takes 4 hours to cook, how long will it take a 14 pound turkey to cook?

3. An 18 ounce box of cereal costs \$2.76. How many ounces should a box priced at \$2.07 contain?

4. Mike and Pat traveled 392 miles in 7 hours. If they travel at the same rate, how long will it take them to travel 728 miles?

5. If 2 pounds of turkey costs \$1.98, what should 3 pounds cost?

6. If 2 liters of fruit juice cost \$3.98, how much do 5 liters cost?

7. A 12 ounce box of cereal costs \$.84. How many ounces should be in a box marked \$.49?

8. Janie saw an advertisement for a 6 ounce tube of toothpaste that costs \$.90. How much should a 4 ounce tube cost?

© Carson-Dellosa • CD-704384

Name________________________________ 6.RP.A.2, 6.RP.A.3b, 7.RP.A.2b, 7.RP.A.2c

Proportions Proverb

Solve each proportion. Place the equation letter above the answer. The statement is a paraphrase of a proverb.

1. $\frac{3}{4} = \frac{12}{A}$
2. $\frac{4}{5} = \frac{C}{15}$
3. $\frac{5}{D} = \frac{20}{10}$
4. $\frac{E}{21} = \frac{2}{6}$
5. $\frac{F}{4} = \frac{7}{8}$
6. $\frac{3}{I} = \frac{30}{7}$
7. $\frac{15}{L} = \frac{20}{L + 0.5}$
8. $\frac{M}{17} = \frac{2}{4}$
9. $\frac{4}{2.8} = \frac{3}{N}$
10. $\frac{P}{3} = \frac{1}{15}$
11. $\frac{Q}{18} = \frac{5}{4}$
12. $\frac{7}{2} = \frac{R}{R-15}$
13. $\frac{5}{S} = \frac{4}{S-2.5}$
14. $\frac{18}{60} = \frac{20}{5T - 1.5}$
15. If four cans of stew cost $5.00, what will be the unit (U) cost of 1 can?
16. Jamie rode her bike 52.5 miles east in 3 hours. What was her velocity (V) per hour?
17. Tony ran for a total of 75 yards during 6 games. At this rate, how many yards (Y) will Tony run in the remaining 2 games?

___ O ___ ___   ___ ___ ___ ___ ___   ___ O ___ ___   ___ ___ ___ ___ ___
2.5   2.1   0.9   0.2   1.5   16   12   7   25   1.25   21   8.5   7   16   2.1   12.5

O ___   ___ O ___ ___ ___ ___ ___ ___ ___ ___   ___ ___
3.5   12   2.1   17.5   7   25   16   2.1   12   7   0.7   2.1

___ ___ O ___ ___   O ___   ___ O ___ ___   ___ ___ ___ ___ ___ ___
3.5   21   2.1   0.9   3.5   25   1.25   21   7   22.5   1.25   0.7   2.1   7

___ ___ ___ ___ ___ ___ .
16   2.1   0.7   8.5   16   1.5

Write the familiar proverb.

________ ________ ________ ________ ________ ________ ________ .

Name________________________________

6.RP.A.3c

Percents

Percent (%) means: per hundred
out of a hundred
hundredths
2 decimal places

$\frac{3}{4} \rightarrow \frac{3}{4} = \frac{x}{100}$

$300 = 4x$

$75 = x$

$\frac{3}{4} = 75\%$

$0.375 \rightarrow 37.5$ hundredths $= 37.5\%$

1. $\frac{4}{5}$
2. $\frac{4}{7}$
3. 0.22
4. 2.5
5. $\frac{3}{8}$
6. 0.006
7. 1.125
8. $\frac{1}{2}$
9. $\frac{9}{40}$
10. 11.3
11. $\frac{11}{20}$
12. 0.086
13. $\frac{7}{8}$
14. 16.688
15. $\frac{7}{16}$
16. 621.9
17. $\frac{5}{16}$
18. 3.9932

© Carson-Dellosa • CD-704384

Name________________________________

6.RP.A.3c

Working with Percents

80% of 30 =

$$\frac{80}{100} = \frac{x}{30}$$

$100x = 2400$

$x = 24$

1. 20% of 10 = ________
2. 25% of 45 = ________
3. 88% of 15 = ________
4. $9\frac{1}{2}$% of 20 = ________
5. 25% of 39 = ________
6. 16% of 90 = ________

___% of 40 = 10

$$\frac{x}{100} = \frac{10}{40}$$

$40x = 1000$

$x = 25$ 25%

1. ________% of 25 = 15
2. ________% of 30 = 10
3. ________% of 4 = 7
4. ________ % of 75 = 33
5. ________ % of 15 = 6
6. ________ % of 80 = 40

50% of ___ = 65

$$\frac{50}{100} = \frac{65}{x}$$

$50x = 6500$

$x = 130$

1. 20% of ________ = 15
2. 80% of ________ = 56
3. 25% of ________ = 19
4. $33\frac{1}{3}$% of ________ = 41
5. 80% of ________ = 16
6. 30% of ________ = 15

© Carson-Dellosa • CD-704384

Name________________________________ 6.RP.A.3c

Problems with Percents

1. In a group of 60 children, 12 have brown eyes. What percent have brown eyes?

2. A salesman makes a 5% commission on all he sells. How much does he have to sell to make $1500?

3. A sales tax of $5\frac{3}{4}$ % is charged on a blouse priced at $42. How much sales tax must be paid?

4. A baby weighed 7.6 pounds at birth and $9\frac{1}{2}$ pounds after 6 weeks. What was the percent increase?

5. A scale model of a building is 8% of actual size. If the model is 1.2 meters tall, how tall is the building?

6. The purchase price of a camera is $84. The carrying case is 12% of the purchase price. Find the total cost including the carrying case.

7. The regular price of a CD is $15. Find the discount and the new price if there is a 20% discount.

8. A basketball team played 45 games. They won 60% of them. How many did the team win?

9. A test had 50 questions. Joe got 70% of them correct. How many did Joe get correct?

10. Diet soda contains 90% less calories than regular soda. If a can of regular soda contains 112 calories, how many calories does a can of diet soda contain?

© Carson-Dellosa • CD-704384

Name________________________

6.RP.A.3c

More Percents

Find 25% of $240.

$$\frac{25}{100} = \frac{x}{240}$$

$$100x = 25 \cdot 240$$

$$100x = 6000$$

$$x = \$60$$

If 20% of a number is 2, find the number.

$$\frac{20}{100} = \frac{32}{x}$$

$$20x = 32 \cdot 100$$

$$20x = 3200$$

$$x = 160$$

The number is 160.

What percent is 15 out of 45?

$$\frac{x}{100} = \frac{15}{45}$$

$$45x = 100 \cdot 15$$

$$45x = 1500$$

$$x = 33\frac{1}{3}$$

$$33\frac{1}{3}\%$$

1. 72% of 310

2. 21 is 35% of what number?

3. 28 out of 70 is what percent?

4. 6% of what number is 2.36?

5. 3.9 is what percent of 10?

6. 115% of 12

7. 60% of what number is 54?

8. 17% of 800 is what number?

9. What percent of 72 is 27?

10. A piece of jewelry costs $78. If the price increases by 12%, what is the new cost?

11. Tax on a $24 item is $1.56. What is the tax rate (percent)?

12. A dress was reduced in price by $19.56. This was 20% of the original price. Find the sale price.

13. There are 252 students on the student council at West High School. If there are 700 students enrolled, what percent are on the student council?

14. One day 3% of the sweatshirts (or 15 sweatshirts) made at a factory were defective. How many sweatshirts were produced at the factory that day?

Name______________________________

Just for Fun

Try to de-code these words and phrases.

STAND

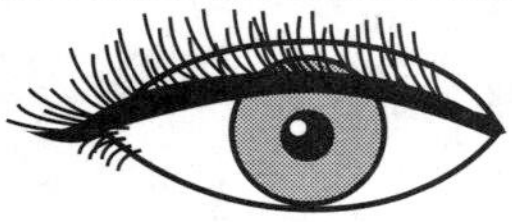

⟶ I understand

1. **sand**	2. **MOMANON**	3. R | E | A | D	4. **WEAR** **LONG**
5. **MCE** **MCE** **MCE**	6. **HANDS** ——— **ACTIVITIES**	7. KEND **VACATION**	8. **FALUTIN**
9. **SIGN** ● ● ● ● ● ●	10. **LET GONES** **GONES** **B GONES** **GONES**	11. **Thought Clever**	12. luck luck LUCK luck *luck* luck luck *luck* luck luck
13. **MAN** **BOARD**	14. 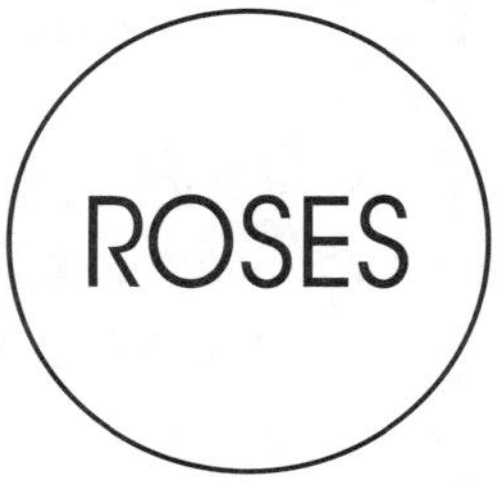	15. 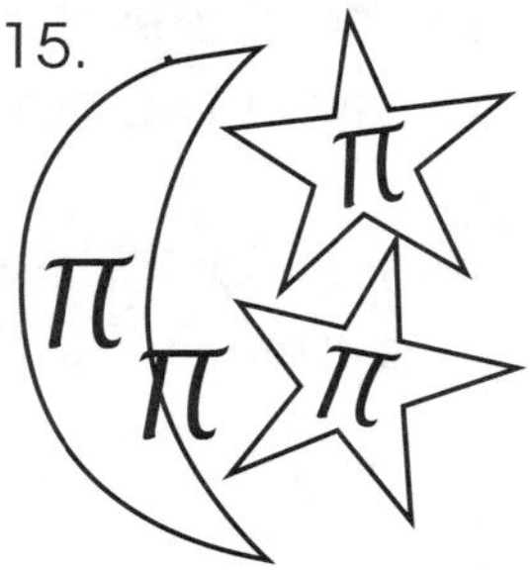	16. **T O U C H**

© Carson-Dellosa • CD-704384

Name________________________________

6.NS.A.1

Adding and Subtracting Fractions

Use the common denominator. Add or subtract the numerators. Reduce to lowest terms.

$\frac{1}{8} + \frac{3}{8} = \frac{4}{8} = \frac{1}{2}$ Add same

1. $\frac{2}{9} + \frac{5}{9} =$

2. $\frac{3}{4} - \frac{1}{4} =$

3. $\frac{9}{15} + \frac{5}{15} =$

4. $\frac{19}{20} - \frac{14}{20} =$

5. $\frac{27}{38} + \frac{13}{38} =$

6. $\frac{35}{60} - \frac{17}{60} =$

7. $\frac{17}{20} + \frac{23}{20} =$

8. $\frac{25}{13} - \frac{12}{13} =$

9. $\frac{11}{18} + \frac{16}{18} =$

10. $\frac{17}{48} - \frac{14}{48} =$

11. $\frac{7}{45} + \frac{8}{45} =$

12. $\frac{33}{50} - \frac{17}{50} =$

13. $\frac{16}{33} + \frac{21}{33} =$

14. $\frac{43}{56} - \frac{19}{56} =$

15. $\frac{12}{42} + \frac{31}{42} =$

16. $\frac{29}{52} - \frac{13}{52} =$

17. $\frac{15}{18} + \frac{8}{18} =$

18. $\frac{43}{65} - \frac{28}{65} =$

© Carson-Dellosa • CD-704384

Name____________________________________

6.NS.A.1

More Adding and Subtracting Fractions

Hint: Remember to rewrite fractions.

$$\frac{7}{9} - \frac{1}{4} = \frac{28}{36} - \frac{9}{36} = \frac{19}{36}$$

Multiply

36 is the least common multiple.

1. $\frac{2}{3} + \frac{5}{9} =$

2. $\frac{4}{5} - \frac{3}{4} =$

3. $\frac{5}{6} + \frac{7}{12} =$

4. $\frac{11}{15} - \frac{2}{5} =$

5. $\frac{11}{12} + \frac{5}{8} =$

6. $\frac{1}{2} - \frac{4}{9} =$

7. $\frac{13}{36} + \frac{5}{12} =$

8. $\frac{7}{8} - \frac{3}{10} =$

9. $\frac{5}{12} - \frac{5}{18} =$

10. $\frac{5}{9} + \frac{3}{8} =$

11. $\frac{5}{12} - \frac{3}{15} =$

12. $\frac{3}{4} + \frac{7}{12} =$

13. $\frac{8}{19} - \frac{1}{3} =$

14. $\frac{7}{15} + \frac{3}{25} =$

15. $\frac{30}{36} - \frac{5}{18} =$

16. $\frac{4}{5} + \frac{12}{13} =$

© Carson-Dellosa • CD-704384

Name________________________________ 6.NS.A.1

Words to the Wise

Write each sum or difference in lowest terms. Cross out the answers below to reveal the "Words to the Wise."

1. $\frac{4}{9} + \frac{13}{15} =$
2. $\frac{5}{6} + \frac{7}{32} =$
3. $\frac{13}{15} - \frac{1}{3} =$
4. $\frac{3}{11} + \frac{6}{7} =$
5. $\frac{5}{9} - \frac{1}{15} =$
6. $\frac{7}{9} + \frac{1}{6} =$
7. $\frac{9}{10} - \frac{3}{20} =$
8. $\frac{11}{42} + \frac{1}{7} =$
9. $\frac{8}{9} - \frac{1}{12} =$
10. $\frac{7}{12} + \frac{31}{42} =$
11. $\frac{11}{12} - \frac{1}{18} =$
12. $\frac{7}{23} - \frac{1}{7} =$
13. $\frac{8}{21} + \frac{36}{49} =$
14. $\frac{7}{9} - \frac{1}{4} =$
15. $\frac{11}{30} + \frac{2}{25} =$
16. $\frac{27}{35} - \frac{11}{30} =$
17. $\frac{7}{8} + \frac{13}{14} =$
18. $\frac{76}{81} - \frac{22}{63} =$
19. $\frac{1}{3} + \frac{2}{3} =$
20. $\frac{23}{45} - \frac{1}{3} =$

$\frac{19}{36}$ CAN	$\frac{17}{24}$ PUT	$\frac{17}{18}$ IT	$\frac{334}{567}$ PLACE	$\frac{7}{8}$ FORTH	1 IS	$1\frac{14}{45}$ WAS	$1\frac{17}{147}$ WHOLE	$\frac{29}{36}$ IT	$\frac{5}{27}$ HALF	$1\frac{5}{96}$ PROPER
$\frac{3}{8}$ THE	$\frac{26}{161}$ THIS	$\frac{17}{42}$ ON	$\frac{31}{36}$ TRY	$1\frac{35}{68}$ EFFORT	$\frac{31}{80}$ AND	$\frac{8}{15}$ SUM	$\frac{9}{30}$ YOU	$\frac{11}{45}$ GET	$\frac{1}{4}$ A	$1\frac{47}{147}$ FRACTION
$\frac{8}{45}$ IF	$\frac{17}{42}$ ARE	$\frac{22}{45}$ IN	$\frac{13}{55}$ OF	$1\frac{10}{77}$ TOTAL	$\frac{67}{150}$ HAS	$\frac{3}{5}$ THE	$1\frac{45}{56}$ VALUE	$1\frac{9}{28}$ TO	$\frac{3}{4}$ ALL	$1\frac{7}{47}$ RESULTS

________ ________ ________ ________ ________ ________

________ ________ ________ ________ ________ ________ ________

© Carson-Dellosa • CD-704384

Name________________________________

6.NS.A.1

Adding and Subtracting Mixed Numbers

add

$$3\frac{7}{8} + 5\frac{11}{24} = 3\frac{21}{24} + 5\frac{11}{24} = 8\frac{32}{24} = 9\frac{8}{24} = 9\frac{1}{3}$$

add

1. $1\frac{1}{4} + 2\frac{1}{2} =$

2. $5\frac{7}{10} - 1\frac{1}{6} =$

3. $8\frac{3}{8} + 9\frac{2}{3} =$

4. $6 - 2\frac{8}{11} =$

5. $2\frac{1}{16} + 2\frac{1}{3} =$

6. $7\frac{7}{8} - 7\frac{5}{12} =$

7. $4\frac{1}{2} + 6\frac{2}{5} =$

8. $5\frac{1}{2} - \frac{11}{15} =$

9. $1\frac{5}{6} + 4 =$

10. $6\frac{7}{9} - 6\frac{1}{2} =$

11. $7\frac{1}{4} + 1\frac{7}{9} + 2\frac{5}{6} =$

12. $8\frac{1}{6} - 7\frac{3}{4} =$

13. $5 + 3\frac{3}{11} =$

14. $3\frac{5}{8} - 1\frac{6}{7} =$

15. $4\frac{3}{7} + 5\frac{5}{14} =$

16. $6\frac{3}{12} - 3\frac{9}{36} =$

© Carson-Dellosa • CD-704384

Name______________________________

6.NS.A.1

Pair Them Up!

Each problem in the first column has the same answer as a problem in the second column. Solve the problems and determine the matches.

1. $7\frac{3}{5} + 2\frac{1}{2} =$

2. $10\frac{3}{5} - 4 =$

3. $5\frac{2}{9} + 7\frac{1}{3} =$

4. $11\frac{5}{6} - 3\frac{3}{4} =$

5. $4\frac{7}{12} + 4\frac{3}{14} =$

6. $8 - 6\frac{5}{9} =$

7. $17\frac{14}{15} + 2\frac{9}{10} =$

8. $1\frac{17}{18} - \frac{1}{8} =$

9. $6\frac{1}{12} + 6\frac{3}{4} =$

10. $8\frac{2}{9} - 6\frac{17}{18} =$

A. $8\frac{1}{3} + 12\frac{1}{2} =$

B. $2\frac{5}{18} - \frac{11}{24} =$

C. $5\frac{2}{9} - 3\frac{7}{9} =$

D. $13 - 6\frac{2}{5} =$

E. $3\frac{5}{6} + 8\frac{13}{18} =$

F. $17\frac{5}{12} - 4\frac{7}{12} =$

G. $7\frac{11}{12} + \frac{1}{6} =$

H. $10\frac{13}{14} - 2\frac{11}{84} =$

I. $5\frac{1}{5} + 4\frac{9}{10} =$

J. $9\frac{1}{6} - 7\frac{8}{9} =$

Name________________________________

6.NS.A.1

Multiplying Fractions

Multiply numerators. Multiply denominators. Reduce to lowest terms.
Hint: Rewrite mixed numbers as improper fractions.

$$2\frac{1}{4} \cdot 1\frac{2}{3} = \frac{9}{4} \cdot \frac{5}{3} = \frac{\overset{3}{\cancel{9}}}{4} \cdot \frac{5}{\underset{1}{\cancel{3}}} = \frac{15}{4} = 3\frac{3}{4}$$

(rewrite $2\frac{1}{4}$ as $\frac{9}{4}$; rewrite $1\frac{2}{3}$ as $\frac{5}{3}$)

1. $\frac{1}{2} \cdot \frac{5}{6} =$

2. $3 \cdot \frac{1}{2} =$

3. $\frac{2}{5} \cdot \frac{1}{3} =$

4. $\frac{16}{5} \cdot \frac{25}{27} =$

5. $\frac{8}{21} \cdot 2\frac{7}{16} =$

6. $1\frac{5}{7} \cdot 2\frac{1}{4} =$

7. $5\frac{7}{8} \cdot 4 =$

8. $\frac{5}{7} \cdot \frac{7}{5} =$

9. $3\frac{2}{3} \cdot \frac{17}{22} =$

10. $\frac{5}{6} \cdot 2 =$

11. $8\frac{1}{3} \cdot \frac{3}{4} =$

12. $4\frac{1}{4} \cdot 3\frac{1}{5} =$

13. $2\frac{1}{6} \cdot \frac{18}{20} =$

14. $\frac{21}{35} \cdot 3\frac{4}{7} =$

15. $1\frac{3}{5} \cdot 2\frac{3}{16} =$

16. $6\frac{3}{4} \cdot 1\frac{5}{9} =$

17. $3\frac{1}{3} \cdot 1\frac{3}{18} =$

18. $\frac{1}{2} \cdot \frac{6}{11} \cdot \frac{3}{5} =$

© Carson-Dellosa • CD-704384

Name________________________________ 6.NS.A.1

Mort's Multiplication

Mort did not understand multiplying mixed numbers when he completed the quiz below. Find and correct the 10 errors Mort made. Explain how to multiply mixed numbers.

FRACTIONS QUIZ

1. $5\frac{3}{5} \cdot 3\frac{4}{7} = 15\frac{12}{35}$
2. $2\frac{1}{12} \cdot 2\frac{2}{15} = 4\frac{4}{9}$
3. $1\frac{1}{15} \cdot 3\frac{3}{7} = 3\frac{23}{35}$
4. $8\frac{2}{9} \cdot 2\frac{7}{8} = 16\frac{7}{36}$
5. $16 \cdot 4\frac{1}{4} = 68$
6. $6\frac{2}{3} \cdot 1\frac{15}{16} = 12\frac{11}{12}$
7. $5\frac{1}{3} \cdot 4\frac{1}{2} = 20\frac{1}{6}$
8. $9\frac{1}{3} \cdot 8\frac{1}{10} = 75\frac{3}{5}$
9. $2\frac{1}{12} \cdot 3\frac{5}{9} = 7\frac{11}{27}$
10. $3\frac{5}{6} \cdot 8 = 24\frac{5}{6}$

Name *Mort*

11. $9\frac{1}{3} \cdot 1\frac{5}{7} \cdot \frac{3}{4} = 9\frac{15}{84}$
12. $6\frac{8}{9} \cdot 3\frac{6}{7} = 26\frac{4}{7}$
13. $8\frac{2}{5} \cdot 3\frac{1}{3} = 24\frac{2}{15}$
14. $9\frac{3}{5} \cdot 2\frac{1}{12} = 20$
15. $2\frac{1}{2} \cdot 2\frac{8}{9} = 4\frac{4}{9}$
16. $5\frac{3}{7} \cdot 2\frac{3}{16} = 10\frac{9}{112}$
17. $2\frac{1}{4} \cdot 6 \cdot 1\frac{1}{9} = 15$
18. $4\frac{1}{2} \cdot 2\frac{2}{5} = 8\frac{1}{5}$
19. $7\frac{1}{2} \cdot 7\frac{1}{3} = 55$
20. $3\frac{1}{8} \cdot \frac{1}{9} \cdot \frac{9}{10} = 3\frac{1}{80}$

© Carson-Dellosa • CD-704384

Name________________________________ 6.NS.A.1

Dividing Fractions

Invert and multiply.

$$1\frac{1}{2} \div 3\frac{3}{7} = \frac{3}{2} \div \frac{24}{7} = \frac{3}{2} \cdot \frac{7}{24} = \frac{\overset{1}{\cancel{3}}}{2} \cdot \frac{7}{\underset{8}{\cancel{24}}} = \frac{7}{16}$$

Rewrite the mixed numbers.

1. $\frac{3}{7} \div \frac{1}{2} =$

2. $\frac{17}{9} \div \frac{8}{9} =$

3. $6\frac{2}{3} \div 5 =$

4. $1\frac{7}{9} \div 4\frac{2}{9} =$

5. $\frac{15}{4} \div \frac{5}{14} =$

6. $\frac{11}{12} \div \frac{13}{8} =$

7. $4 \div 4\frac{2}{5} =$

8. $3\frac{1}{4} \div 4\frac{3}{8} =$

9. $\frac{6}{15} \div \frac{9}{10} =$

10. $\frac{7}{8} \div 2\frac{1}{3} =$

11. $9\frac{3}{8} \div 3\frac{3}{4} =$

12. $5\frac{1}{6} \div \frac{31}{6} =$

13. $\frac{7}{8} \div \frac{3}{4} =$

14. $\frac{7}{12} \div \frac{7}{4} =$

15. $4\frac{6}{7} \div \frac{1}{3} =$

16. $5\frac{1}{2} \div \frac{7}{4} =$

17. $2\frac{2}{9} \div 4\frac{2}{6} =$

18. $5\frac{5}{12} \div 3\frac{1}{3} =$

© Carson-Dellosa • CD-704384

Name________________________________

6.NS.A.1

Division Magic

In a Magic Square, each row, column, and diagonal has the same sum—the Magic Sum. Complete the problems and determine the Magic Sum.

$\frac{5}{12} \div \frac{1}{2}$	$1\frac{1}{2} \div 1\frac{1}{3}$	$\frac{5}{6} \div 5$	$\frac{11}{12} \div 2$	$1\frac{1}{2} \div 2$
$6\frac{1}{2} \div 6$	$\frac{3}{4} \div \frac{9}{4}$	$\frac{5}{6} \div 2$	$2\frac{1}{8} \div 3$	$1\frac{7}{12} \div 2$
$\frac{7}{8} \div 3$	$\frac{6}{7} \div 2\frac{2}{7}$	$\frac{2}{5} \div \frac{3}{5}$	$2\frac{7}{8} \div 3$	$\frac{5}{12} \div \frac{2}{5}$
$\frac{13}{48} \div \frac{1}{2}$	$\frac{5}{32} \div \frac{1}{4}$	$\frac{2}{3} \div \frac{8}{11}$	$\frac{7}{8} \div \frac{7}{8}$	$\frac{1}{2} \div 2$
$1\frac{1}{2} \div 2\frac{4}{7}$	$1\frac{3}{4} \div 2$	$\frac{7}{18} \div \frac{1}{3}$	$\frac{1}{3} \div 1\frac{3}{5}$	$\frac{3}{4} \div 1\frac{1}{2}$

Magic Sum = ______________

If every row and column in a Magic Square of problems has the same sum except for the last row and the last column, what do you know?

____

Name________________________________ 6.NS.A.1

Confused Calculations

Cal Q. Late was very confused about fractions when he completed the quiz below. Find and correct the ten errors Cal made.

FRACTIONS QUIZ

1. $\frac{3}{5} + \frac{1}{3} = \frac{2}{5}$

2. $\frac{3}{4} + \frac{3}{4} = \frac{6}{8}$

3. $4\frac{2}{3} + 6\frac{3}{4} = 10\frac{5}{7}$

4. $2\frac{1}{2} + 3\frac{1}{2} = 6$

5. $\frac{7}{8} - \frac{2}{3} = \frac{5}{24}$

6. $\frac{6}{7} - \frac{2}{7} = \frac{4}{7}$

7. $2\frac{4}{5} - 1\frac{2}{3} = 1\frac{2}{15}$

8. $6\frac{1}{4} - 2\frac{3}{4} = 4\frac{1}{2}$

Name *Cal*

9. $\frac{3}{4} \cdot \frac{6}{7} = \frac{1}{14}$

10. $\frac{1}{3} \cdot \frac{1}{3} = \frac{1}{6}$

11. $1\frac{2}{3} \cdot 2\frac{1}{2} = 2\frac{1}{3}$

12. $4\frac{1}{2} \cdot 3\frac{1}{3} = 12\frac{1}{6}$

13. $\frac{3}{4} \div \frac{1}{2} = 1\frac{1}{2}$

14. $\frac{2}{3} \div \frac{3}{4} = \frac{8}{9}$

15. $2\frac{4}{5} \div 1\frac{2}{5} = 2\frac{2}{5}$

16. $5\frac{1}{4} \div 3\frac{1}{2} = 15\frac{1}{8}$

What rules for computing with fractions would you share with Cal?

Addition ________________________________

Subtraction ________________________________

Multiplication ________________________________

Division ________________________________

© Carson-Dellosa • CD-704384

Name________________________________ 6.NS.A.1

More Mixed Practice with Fractions

1. $8\frac{1}{15} - 5\frac{11}{20} =$

2. $3\frac{1}{9} + 8\frac{3}{7} + 1\frac{1}{3} =$

3. $1\frac{7}{8} \cdot 3\frac{3}{5} =$

4. $4\frac{4}{5} \div 2\frac{8}{10} =$

5. $3\frac{5}{12} + 5\frac{1}{4} - 2\frac{7}{20} =$

6. $(\frac{16}{21} \cdot 3\frac{1}{4}) + 6\frac{1}{3} =$

7. $5\frac{7}{10} - (\frac{25}{27} \div 3\frac{1}{3}) =$

8. $(2\frac{15}{24} + 3\frac{11}{12}) \cdot 6\frac{1}{2} =$

9. $7\frac{3}{12} - 2\frac{8}{9} =$

10. $1\frac{1}{6} \cdot 3\frac{5}{7} \cdot 2\frac{2}{9} =$

11. $8\frac{7}{12} + 11\frac{3}{4} =$

12. $7 - (3\frac{7}{9} \div 4\frac{2}{3}) =$

13. $2\frac{1}{2} \cdot 3\frac{3}{15} =$

14. $5\frac{2}{9} - 2\frac{17}{18} + 1\frac{2}{3} =$

15. $(3\frac{6}{8} \div 4\frac{2}{4}) - \frac{13}{16} =$

16. $4\frac{2}{3} \cdot 1\frac{3}{4} \cdot 3\frac{3}{4} =$

17. $3\frac{4}{15} + 8\frac{3}{45} =$

18. $12\frac{1}{2} - 7\frac{15}{16} =$

19. $(1\frac{12}{13} \cdot 7\frac{3}{5}) - 3 =$

20. $2\frac{1}{8} + (6\frac{2}{3} \div 8\frac{4}{9}) =$

21. $3\frac{1}{3} \cdot 7\frac{5}{6} \cdot 2\frac{2}{5} =$

22. $1\frac{15}{16} + 3\frac{7}{24} + 3\frac{11}{12} =$

© Carson-Dellosa • CD-704384

Name________________________________

6.NS.A.1

Problems with Fractions

1. If $1\frac{1}{4}$ pounds of bananas sell for 80¢ and $1\frac{1}{3}$ pounds of apples sell for 90¢, which fruit is cheaper?

2. A cake recipe calls for $\frac{2}{3}$ teaspoons of salt, $1\frac{1}{2}$ teaspoons baking powder, 1 teaspoon baking soda and $\frac{1}{2}$ teaspoon cinnamon. How many total teaspoons of dry ingredients are used?

3. A baseball team played 35 games and won $\frac{4}{7}$ of them.
   How many games were won?
   How many games were lost?

4. During 4 days, the price of the stock of PEV Corporation went up $\frac{1}{4}$ of a point, down $\frac{1}{3}$ of a point, down $\frac{3}{4}$ of a point, and up $\frac{7}{10}$ of a point. What was the net change?

5. Janie wants to make raisin cookies. She needs $8\frac{1}{2}$ cups of raisins for the cookies. A 15–ounce box of raisins contains $2\frac{3}{4}$ cups. How many boxes must Janie buy to make her cookies?

6. A one–half gallon carton of milk costs $1.89. A one–gallon carton of milk costs $2.99. How much money would you save if you bought a one–gallon carton instead of 2 one–half gallon cartons?

© Carson-Dellosa • CD-704384

Name________________________________

7.NS.A.2d

Changing Fractions to Decimals

$\frac{7}{20} \rightarrow 20\overline{)7.00}$ = .35 → $\frac{7}{20} = 0.35$ terminating

```
     .35
20 ) 7.00
     6.0
     1.00
     1.00
        0
```

$\frac{5}{12} \rightarrow 12\overline{)5.00000}$ = .41666 → $\frac{5}{12} = 0.41\overline{6}$ repeating

```
     .41666
12 ) 5.00000
     4.8
       20
       12
        80
        72
         80
         72
          80
```

1. $\frac{3}{5} =$

2. $\frac{11}{25} =$

3. $\frac{7}{15} =$

4. $2\frac{1}{9} =$

5. $\frac{23}{33} =$

6. $1\frac{5}{16} =$

7. $\frac{12}{25} =$

8. $\frac{1}{3} =$

9. $\frac{5}{33} =$

10. $2\frac{5}{16} =$

11. $\frac{25}{37} =$

12. $3\frac{13}{15} =$

13. $\frac{17}{22} =$

14. $3\frac{11}{12} =$

Name________________________________ 7.NS.A.2d

Terminators

Change each of the following fractions into decimal equivalents. Indicate whether the decimal terminates (T) or repeats (R).

Fraction	Decimal	T or R	Fraction	Decimal	T or R
1. $\frac{3}{8}$			11. $2\frac{3}{8}$		
2. $\frac{8}{15}$			12. $2\frac{15}{37}$		
3. $\frac{27}{32}$			13. $\frac{67}{90}$		
4. $\frac{23}{30}$			14. $1\frac{19}{33}$		
5. $\frac{4}{7}$			15. $\frac{124}{333}$		
6. $5\frac{1}{8}$			16. $5\frac{7}{10}$		
7. $1\frac{4}{5}$			17. $2\frac{11}{16}$		
8. $\frac{10}{35}$			18. $7\frac{31}{40}$		
9. $\frac{9}{15}$			19. $3\frac{9}{16}$		
10. $2\frac{7}{8}$			20. $11\frac{3}{4}$		

BONUS: For fractions in lowest terms, what are the prime factors of the denominators that terminate?

____

Give a rule for determining whether a fraction will be a terminating or repeating decimal.

____

© Carson-Dellosa • CD-704384

Name________________________________

6.NS.B.3

Rounding Decimals

Round 8.135 to the nearest tenth.

8.1<u>3</u>5 → 8.1

(35: less than 5)

Round 32.56713 to the nearest hundredth.

32.56<u>7</u>13 → 32.57

(71: greater than 5)

Round to the nearest whole number.

1. 41.803 =
2. 119.63 =
3. 20.05 =
4. 3.45 =
5. 79.531 =
6. 8.437 =
7. 29.37 =
8. 109.96 =

Round to the nearest tenth.

9. 33.335 =
10. 1.861 =
11. 99.96 =
12. 103.103 =
13. 16.031 =
14. 281.05 =
15. 8.741 =
16. 27.773 =

Round to the nearest hundredth.

17. 69.713 =
18. 5.569 =
19. 609.906 =
20. 247.898 =
21. 5.535 =
22. 67.1951 =
23. 14.0305 =
24. 6.9372 =

Name________________________________

6.NS.B.3

Multiplying and Dividing by 10, 100, etc.

When multiplying by a power of 10, move the decimal to the right:

$34.61 \times 1\underline{0} \rightarrow$ move 1 place $\rightarrow 346.1$

$6.77 \times 1\underline{00} \rightarrow$ move 2 places $\rightarrow 677$

When dividing by a power of 10, move the decimal to the left:

$7.39 \div 1\underline{00} \rightarrow$ move 2 places $\rightarrow 0.0739$

$105.61 \div 1\underline{000} \rightarrow$ move 3 places $\rightarrow 0.10561$

1. 4.81 x 100 =
2. 37.68 ÷ 10 =
3. 0.46 x 1,000 =
4. 7.12 ÷ 10,000 =
5. 5.4 x 10 =
6. 27,500 ÷ 1,000 =
7. 4.395 x 100,000 =
8. 0.0075 ÷ 100 =
9. 2.274 x 10 =
10. 90,000 ÷ 100 =
11. 0.000618 x 1,000 =
12. 39.006 ÷ 1,000 =
13. 16 x 100 =
14. 28.889 ÷ 10,000 =
15. 36.89 x 10,000 =
16. 0.091 ÷ 100 =
17. 0.0336 x 100,000 =
18. 1,672 ÷ 100,000 =

© Carson-Dellosa • CD-704384

Name________________________________

6.NS.B.3

Adding and Subtracting Decimals

13.6 + 7.12 =
```
  13.6
+ 7.12
------
 20.72
```

12 − 3.78 =
```
  12
− 3.78
------
  8.22
```

1. 3.5 + 8.4 =

2. 43.57 + 104.6 =

3. 15.36 + 29.23 + 7.2 =

4. 2.304 + 6.18 + 9.2 =

5. $12.91 + $6.99 =

6. 0.08 + 19 =

7. 22.63 + 1.694 =

8. 362.1 + 8.888 + 0.016 =

9. 1392.16 + 16.16 =

10. 83.196 + 0.0017 =

11. 17.6 − 9.3 =

12. 32.3 − 12.72 =

13. 23.96 − 19.931 =

14. $29.98 − $16.09 =

15. 63.36 − 0.007 =

16. 16.22 − 0.039 =

17. 44.44 − 16.1 =

18. $75.02 − $3.99 =

19. 575.021 − 65.98 =

20. 394.6 − 27.88 − 0.0933 =

Name____________________________________ 6.NS.B.3

More or Less

Compute the sums and differences. Cross out each answer below. The remaining letters spell out an important rule.

1. 6.2 + 0.25 = __________
2. 3.3 − 0.33 = __________
3. 0.26 + 0.4 = __________
4. 8.76 − 5.43 = __________
5. 19.9 + 1.1 = __________
6. 9.53 − 5.3 = __________
7. 0.22 + 2.2 = __________
8. 77.7 − 7 = __________
9. 7.8 + 64.2 = __________
10. 9.25 − 2.5 = __________
11. 36 + 6.3 = __________
12. 37.2 − 32 = __________
13. 0.23 + 3.7 = __________
14. 28.55 − 20.5 = __________
15. 27.8 + 2.2 − 3.5 + 0.5 − 20.5 = __________

20 RE	3.33 AL	34 ME	70.7 WA	71 MB	6.75 YS	0.3 ER	6.45 CO	0.66 UN	9 TO	42.3 AD	3.93 DU	8.7 LI	8.05 PT
9.9 NE	2.42 HE	4.4 UP	5.2 PL	4.49 TH	77 EP	72 AC	0.6 OI	21 ES	4.23 TO	9 NT	2.97 AD	6.5 D.	3.07 S.

Write the remaining letters, one letter to a space.

__ __ __ __ __ __ __ __   __ __

__ __ __ __   __ __   __ __ __

__ __ __ __ __ __

© Carson-Dellosa • CD-704384

Name________________________________

6.NS.B.3

Multiplying Decimals

The number of decimal places in a product equals the sum of decimal places in the factors.

(0.7)	(0.04)	=	0.028
1 +	2	=	3
place	places		places

1. (0.003) (6) =

2. (0.051) (0.003) =

3. (260) (0.01) =

4. (9.6) (5) =

5. (7) (3.42) =

6. (5.29) (11.3) =

7. (0.017) (6.2) =

8. (0.3) (0.03) (0.003) =

9. (1.5) (0.096) (4.3) =

10. (0.05) (0.16) (0.001) =

11. (8) (0.217) (0.01) =

12. (18) (0.08) =

13. (16.01) (0.5) (0.31) =

14. (1.06) (0.005) =

15. (4.802) (11.11) =

16. (10.25) (0.331) =

17. (5) (1.102) =

18. (12.8) (0.05) (3.09) =

Name________________________________

6.NS.B.3

Get to the Point

For each multiplication problem, locate the decimal point in the product. Insert zeros if needed.

1. $2.2 \times 0.011 = 242$
2. $12.8 \times 0.12 = 1536$
3. $1.8 \times 6.03 = 10854$
4. $34.1 \times 1.4 = 47.74$
5. $7.21 \times 22.2 = 160062$
6. $55 \times 0.033 = 1815$
7. $6.9 \times 11 = 759$
8. $6.7 \times 0.801 = 5.3667$
9. $4.04 \times 4.04 = 163216$
10. $32.1 \times 2.02 = 64842$
11. $0.005 \times 0.011 = 55$
12. $66.2 \times 1.1 = 7282$
13. $0.84 \times 0.07 = 588$
14. $8.2 \times 0.1 = 82$
15. $0.6 \times 1.7 = 102$
16. (5.7) (0.2) (0.07) = 798
17. (9.8) (2.8) (1.8) = 49392
18. (10.6) (4.3) (0.8) = 36464
19. (0.13) (8.5) (0.5) = 5525
20. (6.7) (1.2) (0.03) = 2412

HINT:
The sum of the number of all decimal places in your products equals 64.

© Carson-Dellosa • CD-704384

Name______________________________

6.NS.B.3

Dividing Decimals

HINT:
Move the decimal points the number of places needed to make the divisor a whole number.

$$0.03652 \div .88 = \quad .88 \overline{)\,.036520} \quad \text{quotient: } .0415$$

```
       .0415
.88 ⟌ .036520
       352
       132
        88
        440
        440
          0
```

1. $0.128 \div 0.8 =$

2. $2.45 \div 3.5 =$

3. $0.5773 \div 5.02 =$

4. $39.78 \div 0.195 =$

5. $4.2016 \div 5.2 =$

6. $1.45 \div 0.08 =$

7. $0.1716 \div 5.2 =$

8. $3.906 \div 1.2 =$

9. $6.56 \div 0.16 =$

10. $0.0135 \div 4.5 =$

11. $0.0483 \div 0.21 =$

12. $0.5418 \div 0.3 =$

13. $16.83 \div 0.11 =$

14. $0.1926 \div 32.1 =$

Name________________________________ 6.NS.B.3

Mixed Practice with Decimals

1. 12.16 – 8.72 =

2. 119.7 + 11.97 =

3. (3.4) (8) =

4. 2960 ÷ 0.37 =

5. 1.21 ÷ 1.1 =

6. 7 + 6.91 =

7. 18.91 – 11.857 =

8. (1.35) (21.4) =

9. 21.2 – 9.03 =

10. 0.7 + 0.02 + 4 =

11. (0.25) (2.5) (25) =

12. 95.6 – 87.81 +12.21 =

13. (0.8) (1.3) (0.62) =

14. 37.92 ÷ 1.2 =

15. 0.1007 ÷ 5.3 =

16. 329.82 + 6.129 =

17. 893.631 – 11.09 =

18. 18.332 + 82.82 =

19. 132.03 ÷ 8.1 =

20. (16.1) (3.66) =

21. 1093.62 – 10.993 =

22. 6.963 ÷ 2.11 =

© Carson-Dellosa • CD-704384

Name________________________________

6.NS.B.3

Going Around the Block

Start at 0.5. Move clockwise. Fill the blank spaces with +, –, x , or ÷ to make true math statements. End back at 0.5.

Start ⇒

0.5		0.2	=	0.1		0.5	=	0.2		0.8	=	0.25
=												
0.3												0.75
												=
0.15												1
=												
0.1												0.6
												=
1.5												0.4
=												
3												0.3
												=
4.5	=	2.5		2	=	1.4		0.6	=	5		0.12

Name________________________________

6.NS.B.3

Problems with Decimals

1. Jim's gas credit card bill was $80.97 for June, $41.35 for July, and $65.08 for August. What were his total charges for the summer?

2. One cup of hot chocolate can be made with .18 ounces of hot chocolate mix. How many cups can be made from a 6.48 ounce canister of mix?

3. Karl's car payments are $215.37 per month for the next three years. What will be the total amount he will pay for his car?

4. The dress Sally wants costs $85.15. If the price was reduced by $12.78, how much will she pay?

5. Melissa went to the mall and noticed that the price of a coat she wanted was cut in half! The original price was $58.22. What is the sales price?

6. Tyler decided that he wanted a dog. He went to the pet store and bought one for $42.95. Tyler also bought three bags of food for $12.55 a bag. How much did Tyler spend altogether?

7. Christopher decided to make his grandmother a birdhouse instead of buying her one. The materials for the birdhouse totaled $21.99. The cost of a new birdhouse is $37.23. How much did Christopher save?

8. Jim thought that snow skiing looks like lots of fun. He decided to try it. First, he needs equipment. He bought a pair of skis for $129.78, a pair of boots for $62.22, poles for $12.95, a hat for $2.50, a coat for $49.95, ski pants for $27.50, and gloves for $11.25. How much did Jim spend altogether?

© Carson-Dellosa • CD-704384

Name________________________________

7.NS.A.2d

Changing Decimals to Fractions

Terminating Decimals

$0.25 = \frac{25}{100} = \frac{1}{4}$

$0.132 = \frac{132}{1000} = \frac{33}{250}$

Repeating Decimals

$N = 0.\overline{12} = 0.121212...$

$100N = 12.1212...$

$^{-}N = {}^{-}0.1212...$

$\frac{99N}{99} = \frac{12}{99}$

$N = \frac{4}{33}$

or $0.\overline{12} = \frac{4}{33}$

1. $0.125 =$

2. $0.\overline{6} =$

3. $0.36 =$

4. $0.\overline{46} =$

5. $0.6875 =$

6. $0.91\overline{6} =$

7. $0.625 =$

8. $0.\overline{27} =$

9. $0.3\overline{8} =$

10. $0.55 =$

11. $0.5625 =$

12. $0.775 =$

Name________________________________

6.NS.B.3

Can You Decode the Puzzle?

Decipher the code and perform the indicated operations.

.3	$\frac{1}{20}$	2.1
3.1	2.8	$\frac{8}{25}$
4	.1	$\frac{1}{2}$

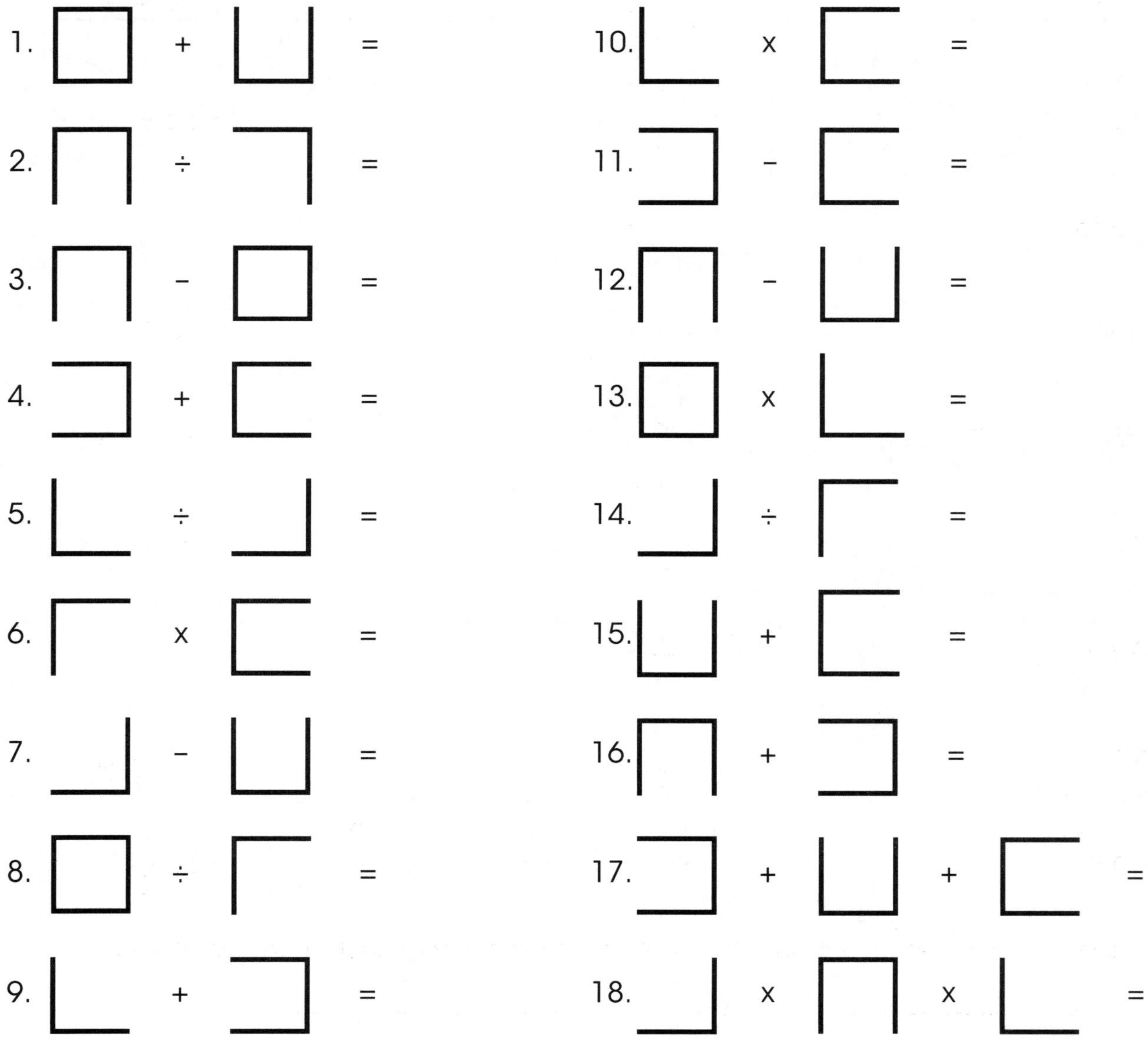

© Carson-Dellosa • CD-704384

Name________________________________ 7.NS.A.2d

Triple Match

Use a ruler to connect each decimal to its fraction equivalent. Then, draw a line connecting the fraction to its percent equivalent. Each path (decimal → fraction → percent) will pass through a letter and a number. Write the letter on the blank above the corresponding number at the bottom of the page.

$0.\overline{33}$ •		• $\frac{3}{8}$ •	10	• 25 %
0.25 •	H N	• $\frac{3}{4}$ •	9	• $37\frac{1}{2}$ %
0.75 •	S	• $\frac{1}{8}$ •		• 75 %
0.125 •	A	• $\frac{5}{6}$ •	11 2	• $12\frac{1}{2}$ %
0.375 •	P	• $\frac{1}{3}$ •	8	• $33\frac{1}{3}$ %
$0.\overline{66}$ •	E	• $\frac{1}{4}$ •	7 4	• 40 %
$0.8\overline{3}$ •	R Y	• $\frac{2}{5}$ •		• $16\frac{2}{3}$ %
0.4 •		• $\frac{1}{6}$ •	1	• 50 %
0.7 •	O	• $\frac{7}{10}$ •	3	• 70 %
0.5 •	g	• $\frac{2}{3}$ •	6	• $83\frac{1}{3}$ %
$0.1\overline{6}$ •	T	• $\frac{1}{2}$ •	5	• $66\frac{2}{3}$ %

Secret society of mathematicians that studied geometric ratios such as the golden ratio:

___ ___ ___ ___ ___ ___ ___ ___ ___ ___ ___ ___

2 4 6 8 10 1 3 5 7 10 9 11

Name______________________________ | 6.NS.B.3

Flip Trip

Perform each of the following operations on your calculator. Then, flip your calculator and find the "word answer" to the questions.

1. What did Amelia Earhart's father say the first time he saw her fly an airplane?

   0.115 x 3 + 10141 x 5 = ______________________________

   Flip Trip ______________________________

2. What did Farmer Macgregor throw at Peter Rabbit to chase him out of the garden?

   (27 x 109 + 4 – 0.027) 2 x 9 = ______________________________

   Flip Trip ______________________________

3. What did Snoopy add to his doghouse as a result of his dogfights with the Red Baron?

   7 (3 x 303 + 50) x 8 = ______________________________

   Flip Trip ______________________________

4. What kind of double does a golfer want to avoid at the end of a round of golf?

   4 (1956 x 4 +153) = ______________________________

   Flip Trip ______________________________

5. What did the little girl say when she was frightened by the ghost?

   0.07 x 0.111 x 5 + 0.00123 = ______________________________

   Flip Trip ______________________________

© Carson-Dellosa • CD-704384

Name________________________________

6.NS.C.6a, 7.NS.A.1b, 7.NS.A.1d

Adding Integers (Number Line)

Label each arrow. Write the resulting expression.

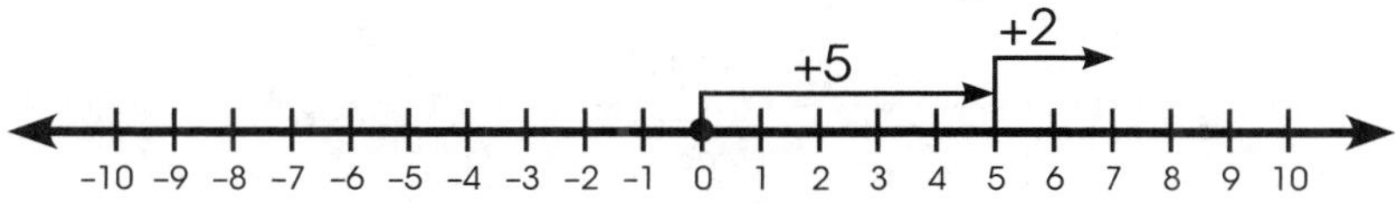

$5 + 2 = 7$

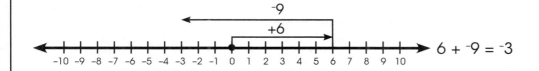

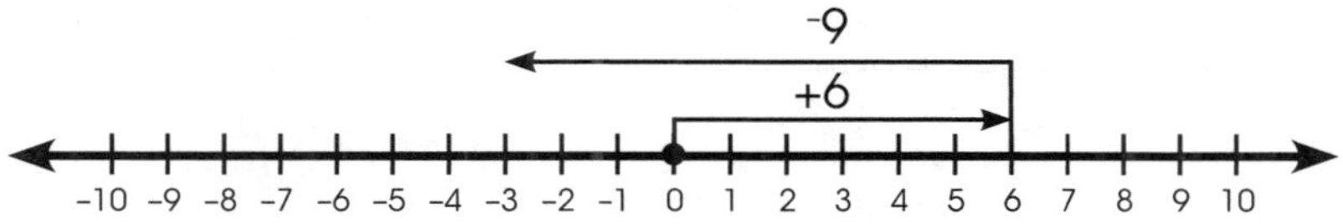

$6 + {}^{-}9 = {}^{-}3$

1. 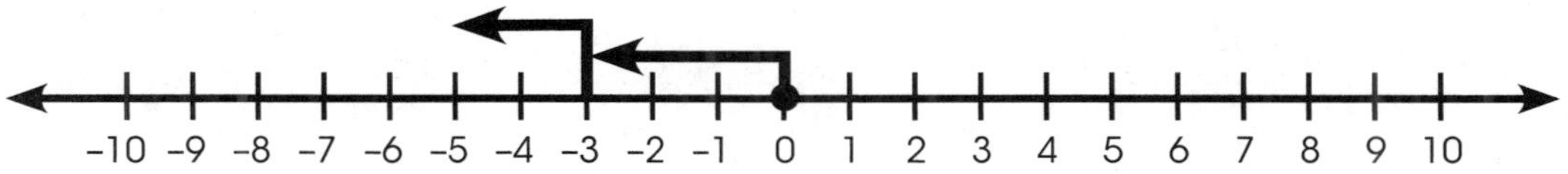

2. 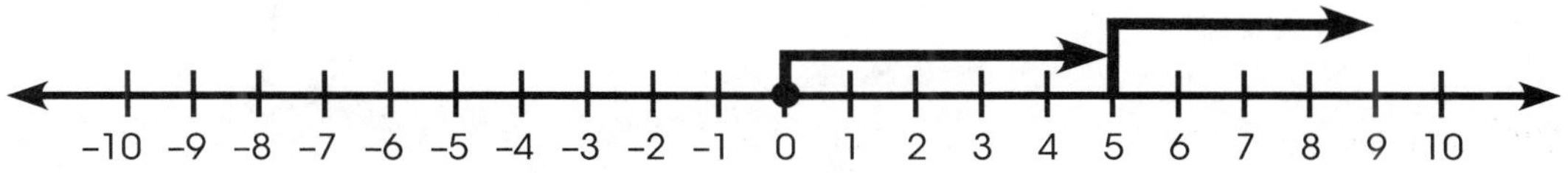

3. 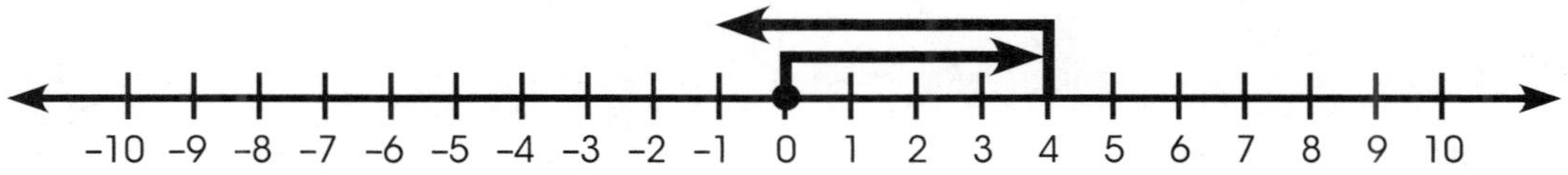

4. 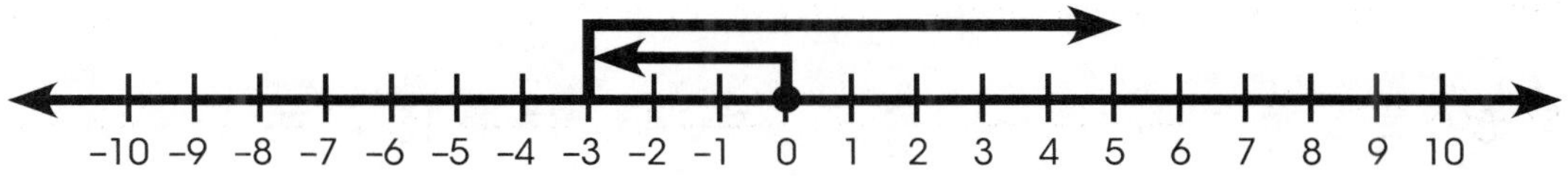

5. 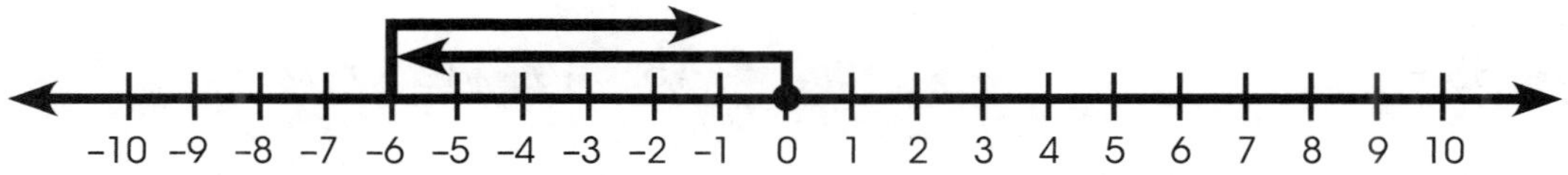

6. 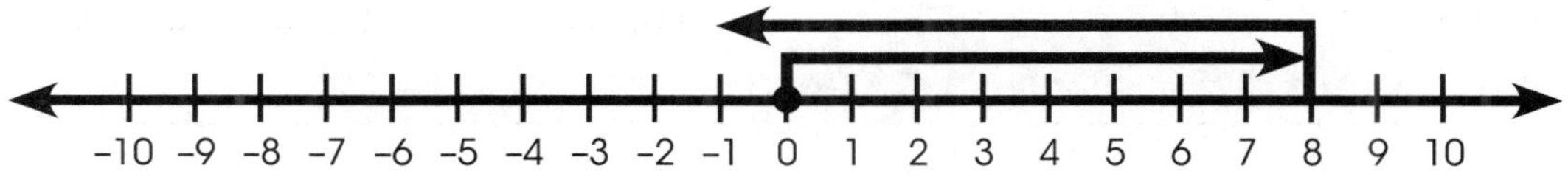

Name________________________________

7.NS.A.1b, 7.NS.A.1d, 7.NS.A.3

Adding Integers with Like Signs

5 + 5	=	10
2 positives		positive
$^-3 + {}^-12$	=	$^-15$
2 negatives		negative

1. $6 + 8 =$

2. $^-9 + {}^-23 =$

3. $25 + 37 =$

4. $^-85 + {}^-19 =$

5. $132 + 899 =$

6. $^-104 + {}^-597 =$

7. $^-642 + {}^-33 =$

8. $88 + 298 =$

9. $^-45 + {}^-68 =$

10. $^-12 + {}^-18 + {}^-35 =$

11. $21 + 108 + 111 =$

12. $^-62 + {}^-33 + {}^-12 =$

13. $17 + 39 + 44 =$

14. $^-18 + {}^-18 + {}^-18 =$

15. $19 + 42 + 647 =$

16. $^-29 + {}^-108 + {}^-337 + {}^-503 =$

© Carson-Dellosa • CD-704384

Name______________________________ 7.NS.A.1b, 7.NS.A.1d, 7.NS.A.3

Adding Integers with Unlike Signs

To add integers with different signs:

Use the sign of the number farther from zero.

Find the difference of the two numbers.

(sign) → ↓	(sign) → ↓
$18 + {}^{-}23 = {}^{-}5$	${}^{-}16 + 19 = +3$
$(23 - 18)$ → → ↑	$(19 - 16)$ → → ↑

1. $21 + {}^{-}87 =$
2. ${}^{-}63 + 59 =$
3. $12 + {}^{-}12 =$
4. ${}^{-}28 + 82 =$
5. ${}^{-}32 + 97 =$
6. ${}^{-}53 + 74 =$
7. $132 + {}^{-}87 =$
8. $212 + {}^{-}99 =$
9. ${}^{-}331 + 155 =$
10. ${}^{-}413 + 521 =$
11. $8{,}129 + {}^{-}6{,}312 =$
12. ${}^{-}11{,}332 + 566 =$
13. $1{,}627 + {}^{-}7{,}193 =$
14. $7{,}864 + {}^{-}6{,}329 =$
15. ${}^{-}10{,}822 + 6{,}635 =$
16. $13{,}894 + {}^{-}81{,}139 =$
17. ${}^{-}16{,}742 + 65{,}524 =$
18. ${}^{-}56{,}814 + 73{,}322 =$
19. $101{,}811 + {}^{-}322{,}885 =$
20. $562{,}493 + {}^{-}112{,}819 =$
21. $116{,}667 + {}^{-}912{,}182 =$
22. ${}^{-}629{,}922 + 81{,}962 =$
23. ${}^{-}196{,}322 + 422{,}899 =$
24. $467{,}833 + {}^{-}36{,}838 =$

© Carson-Dellosa • CD-704384

Name________________________________ 7.NS.A.1b, 7.NS.A.1d, 7.NS.A.3

Integer Grid

Fill in the blanks so that the last number of each row is the sum of the numbers in that row and the bottom number of each column is the sum of the numbers in that column.

3	-1	5		-3		0	4	-8	
2	6	0	-4	-8	2	-7	1		-3
-9		-8	1	4	7		-3	6	2
4	-8	1	-5	9	-6	2	-6	0	
-3		2	-6		7	-1		9	8
5	-8	1	-4	7		-5	9	-2	2
	0		3	-7	1	5	9	2	0
3	-7	4	-8		6	0	4	-9	-5
5	8	-2	6			6	-9	-2	9
	2	-4	-8		6	0	4		6

© Carson-Dellosa • CD-704384

Name________________________________ 7.NS.A.1c, 7.NS.A.1d, 7.NS.A.3

Subtracting Integers

Re-write each problem as an addition problem and solve.

$6 - 11 = 6 + {}^{-}11 = {}^{-}5$ (add the opposite)

$26 - {}^{-}67 = 26 + 67 = 93$ (add the opposite)

1. $19 - 23 =$

2. ${}^{-}8 - 7 =$

3. $35 - 20 =$

4. ${}^{-}46 - {}^{-}18 =$

5. ${}^{-}118 - 12 =$

6. $7 - {}^{-}103 =$

7. $211 - 108 =$

8. ${}^{-}9 - {}^{-}16 =$

9. $63 - 72 =$

10. ${}^{-}93 - 117 =$

11. $45 - {}^{-}50 =$

12. ${}^{-}18 - {}^{-}12 =$

13. $21 - 82 =$

14. ${}^{-}831 - 616 =$

15. ${}^{-}632 - {}^{-}714 =$

16. $1{,}192 - {}^{-}983 =$

Name________________________________

7.NS.A.1c, 7.NS.A.1d, 7.NS.A.3

More Subtracting Integers

1. $7 - 13 =$

2. $^{-}17 - 9 =$

3. $^{-}11 - 7 =$

4. $^{-}24 - {}^{-}23 =$

5. $2 - 25 =$

6. $0 - {}^{-}14 =$

7. $^{-}3 - {}^{-}7 =$

8. $^{-}8 - {}^{-}27 =$

9. $^{-}29 - 36 =$

10. $^{-}72 - {}^{-}84 =$

11. $63 - 94 =$

12. $77 - {}^{-}27 =$

13. $^{-}23 - {}^{-}96 =$

14. $^{-}70 - 18 =$

15. $318 - {}^{-}864 =$

16. $^{-}626 - 118 =$

17. $553 - {}^{-}764 =$

18. $^{-}832 - 1{,}129 =$

19. $6{,}793 - {}^{-}8{,}329 =$

20. $^{-}7{,}624 - 11{,}652 =$

21. $108{,}719 - {}^{-}96{,}989 =$

22. $^{-}832{,}629 - {}^{-}163{,}864 =$

23. $^{-}629{,}299 - 532{,}106 =$

24. $735{,}300 - {}^{-}800{,}919 =$

© Carson-Dellosa • CD-704384

Name________________________ 7.NS.A.1b, 7.NS.A.1c, 7.NS.A.1d, 7.NS.A.3

Adding and Subtracting Integers

1. $^{-}6 + {}^{-}8 =$

2. $^{-}10 - 3 =$

3. $^{-}14 + 20 =$

4. $31 - {}^{-}9 =$

5. $^{-}17 + 9 =$

6. $^{-}8 - {}^{-}27 =$

7. $^{-}33 - 36 =$

8. $19 + {}^{-}32 =$

9. $112 - {}^{-}52 =$

10. $8 - {}^{-}7 =$

11. $24 + {}^{-}24 =$

12. $508 - 678 =$

13. $^{-}23 - {}^{-}28 =$

14. $0 - 31 =$

15. $^{-}40 - 35 =$

16. $73 + {}^{-}19 =$

17. $^{-}231 - {}^{-}231 =$

18. $^{-}107 + {}^{-}293 =$

19. $52 + {}^{-}41 - 60 =$

20. $^{-}85 - {}^{-}106 + 18 =$

21. $81 - 165 - {}^{-}75 =$

22. $^{-}16 + 312 + {}^{-}621 =$

23. $^{-}121 + {}^{-}632 - {}^{-}11 =$

24. $^{-}553 - {}^{-}632 + {}^{-}85 =$

© Carson-Dellosa • CD-704384

Name________________________________ 7.NS.A.2a, 7.NS.A.2c, 7.NS.A.3

Multiplying Integers

$(4)\ (4) = 16$ — $+ \cdot + = +$

$(^-8)\ (^-6) = 48$ — $- \cdot - = +$

Like Signs ⟹ Positive

$(^-5)\ (10) = {}^-50$ — $- \cdot + = -$

Unlike Signs ⟹ Negative

1. $(^-3)\ (^-6) =$
2. $(14)\ (^-4) =$
3. $(25)\ (2) =$
4. $(20)\ (^-49) =$
5. $(75)\ (15) =$
6. $(^-30)\ (^-30) =$
7. $(^-17)\ (23) =$
8. $(^-218)\ (^-32) =$
9. $(801)\ (^-37) =$
10. $(^-89)\ (^-321) =$
11. $(31)\ (^-31)\ (31) =$
12. $(^-4)\ (^-18)\ (28) =$
13. $(^-53)\ (^-14)\ (^-7) =$
14. $(32)\ (125)\ (11) =$
15. $(^-37)\ (^-18)\ (^-5)\ (2) =$
16. $(111)\ (^-63)\ (19) =$
17. $(20)\ (^-7)\ (35)\ (^-3) =$
18. $(16)\ (^-8)\ (^-10)\ (^-1) =$
19. $(^-9)\ (^-29)\ (32)\ (2) =$
20. $(^-18)\ (^-6)\ (^-21)\ (^-30) =$

© Carson-Dellosa • CD-704384

Name________________________________

7.NS.A.2a, 7.NS.A.2c, 7.NS.A.3

Multiplying Real Numbers

$2 \cdot {}^{-}4 = {}^{-}8$ $\quad (-\frac{1}{2})(-\frac{3}{4}) = \frac{3}{8}$

1. $^{-}4 \cdot 15 =$
2. $(^{-}6)\ (^{-}8) =$
3. $(^{-}10)\ (^{-}3)\ (4) =$
4. $(^{-}21)\ (^{-}4)\ (0) =$
5. $(^{-}3)\ (^{-}3)\ (^{-}3) =$
6. $14\ (^{-}6) =$
7. $^{-}40 \times {}^{-}9 =$
8. $(4)\ (^{-}2)\ (3)\ (^{-}1)\ (5)\ (^{-}6) =$
9. $(^{-}1)\ (^{-}1)\ (^{-}1)\ (^{-}1)\ (^{-}1)\ (^{-}1) =$
10. $(1.2)\ (^{-}5) =$
11. $(6.5)\ (^{-}1)\ (^{-}3) =$
12. $(-\frac{5}{8})\ (-\frac{2}{3}) =$
13. $(\frac{3}{8})\ (\frac{5}{6}) =$
14. $(^{-}12)\ (-\frac{1}{3})\ (\frac{3}{4}) =$
15. $(^{-}6\frac{2}{3})\ (3\frac{3}{4}) =$
16. $(^{-}2)^3 =$
17. $(^{-}3)^2 =$
18. $(^{-}1)^{99} =$

Complete the statements with either a positive or negative.

19. A problem with an even number of negative factors will have a ________________ product.
20. A problem with an odd number of negative factors will have a ________________ product.

Name________________________________

7.NS.A.2b, 7.NS.A.2c, 7.NS.A.3

Dividing Integers

$\frac{^{-}24}{^{-}8} = 3$

$\frac{-}{-} = +$

Like Signs ⟹ Positive

$^{-}32 \div 4 = {^{-}8}$

$- \div + = -$

Unlike Signs ⟹ Negative

1. $^{-}49 \div 7 =$

2. $100 \div {^{-}4} =$

3. $^{-}75 \div {^{-}15} =$

4. $^{-}84 \div 21 =$

5. $^{-}120 \div 5 =$

6. $57 \div {^{-}19} =$

7. $^{-}288 \div {^{-}4} =$

8. $804 \div 67 =$

9. $\frac{17}{^{-}17} =$

10. $\frac{^{-}72}{^{-}18} =$

11. $\frac{^{-}195}{13} =$

12. $\frac{^{-}23}{^{-}1} =$

13. $\frac{200}{10} =$

14. $\frac{270}{^{-}45} =$

15. $\frac{^{-}343}{7} =$

16. $\frac{^{-}1125}{-45} =$

© Carson-Dellosa • CD-704384

Name____________________________________

7.NS.A.2b, 7.NS.A.2c, 7.NS.A.3

Dividing Real Numbers

$^{-}5.4 \div {}^{-}9 = 0.6$

1. $^{-}91 \div 7 =$

2. $36 \div ({}^{-}9) =$

3. $^{-}54 \div ({}^{-}9) =$

4. $75 \div 15 =$

5. $0 \div ({}^{-}7) =$

6. $\frac{56}{^{-}7} =$

7. $\frac{^{-}72}{^{-}12} =$

8. $\frac{102}{^{-}17} =$

9. $600 \div 24 =$

10. $\frac{144}{^{-}12} =$

11. $^{-}48 \div 3 =$

12. $^{-}1.5 \div ({}^{-}3) =$

13. $2.4 \div ({}^{-}1.2) =$

14. $^{-}1.44 \div (.3) =$

15. $\frac{0}{^{-}4.12} =$

16. $\frac{1}{8} \div -\frac{6}{5} =$

17. $-\frac{3}{7} \div -\frac{8}{21} =$

18. $^{-}10 \div \frac{1}{3} =$

19. $-\frac{3}{4} \div ({}^{-}12) =$

20. $^{-}15 \div \frac{3}{5} =$

21. $\frac{4}{5} \div (-\frac{3}{10}) =$

22. $-\frac{3}{8} \div (-\frac{3}{4}) =$

23. $\frac{5}{6} \div \frac{4}{9} =$

24. $^{-}6\,\frac{2}{3} \div 3\,\frac{3}{4} =$

© Carson-Dellosa • CD-704384

Name________________________________ 7.NS.A.2b, 7.NS.A.2c, 7.NS.A.3

Divide and Conquer

Compute. Substitute the values into the problem below.

A. $^-81 \div {}^-9 =$

B. $13 \div {}^-13 =$

C. $^-60 \div 10 =$

D. $^-88 \div {}^-11 =$

E. $^-104 \div 8 =$

F. $^-147 \div {}^-21 =$

G. $80 \div {}^-5 =$

H. $52 \div 4 =$

I. $^-150 \div {}^-6 =$

J. $\frac{^-102}{17} =$

K. $\frac{^-75}{^-5} =$

L. $\frac{196}{^-14} =$

M. $\frac{1378}{^-26} =$

N. $\frac{^-468}{^-26} =$

O. $\frac{253}{11} =$

P. $\frac{^-465}{^-31} =$

Q. $\frac{^-552}{^-23} =$

R. $\frac{^-1824}{^-48} =$

William I of Normandy conquered England in → → → ↴

(A+B+C+D+E+F) – (G+H) – [I÷(J+K+L)] – M•N + (O+P+Q+R) = 1066

(_ + _ + _ + _ + _ + _) – (_ + _) – [_ ÷ (_ + _ + _)] – _ • _ + (_ + _ + _ + _)

© Carson-Dellosa • CD-704384

Name______________________________

7.NS.A.1d, 7.NS.A.2c, 7.NS.A.3

Mixed Practice with Integers

1. $^{-}41 + {}^{-}125 =$

2. $79 - 88 =$

3. $^{-}3 \cdot {}^{-}4 =$

4. $\frac{^{-}125}{5} =$

5. $19 \cdot {}^{-}24 =$

6. $\frac{^{-}123}{41} =$

7. $82 + {}^{-}95 =$

8. $27 - {}^{-}46 =$

9. $^{-}31 - {}^{-}32 =$

10. $\frac{^{-}825}{^{-}33} =$

11. $^{-}34 + 52 + {}^{-}18 =$

12. $14 \cdot {}^{-}12 \cdot 3 =$

13. $\frac{^{-}185}{5} \cdot {}^{-}4 =$

14. $76 - 19 + {}^{-}60 =$

15. $17 - {}^{-}12 - 22 =$

16. $100 \cdot {}^{-}4 \cdot 40 =$

17. $\frac{54}{^{-}9} + \frac{33}{11} + \frac{24}{8} =$

18. $^{-}51 \div 17 =$

19. $4 - 8 + {}^{-}9 =$

20. $-\frac{98}{49} \cdot {}^{-}10 =$

21. $(256 \div {}^{-}16) \cdot {}^{-}3 =$

22. $({}^{-}18 - {}^{-}26 + {}^{-}13) \cdot {}^{-}2 =$

23. $(202 + {}^{-}196 - 321) \div {}^{-}5 =$

24. $(\frac{^{-}575}{23} - 18) \cdot {}^{-}11 =$

© Carson-Dellosa • CD-704384

Name_______________________________ 6.NS.C.5, 7.NS.A.1d, 7.NS.A.2c, 7.NS.A.3

Problems with Integers

1. An elevator started at the first floor and went up 18 floors. It then came down 11 floors and went back up 16. At what floor was it stopped?

2. At midnight, the temperature was 30°F. By 6:00 a.m., it had dropped 5° and by noon, it had increased by 11°. What was the temperature at noon?

3. Some number added to 5 is equal to $^{-}11$. Find the number.

4. From the top of a mountain to the floor of the valley below is 4,392 feet. If the valley is 93 feet below sea level, what is the height of the mountain?

5. During one week, the stock market did the following: Monday rose 18 points, Tuesday rose 31 points, Wednesday dropped 5 points, Thursday rose 27 points, and Friday dropped 38 points. If it started out at 1,196 on Monday, what did it end up on Friday?

6. An airplane started at 0 feet. It rose 21,000 feet at takeoff. It then descended 4,329 feet because of clouds. An oncoming plane was approaching, so it rose 6,333 feet. After the oncoming plane passed, it descended 8,453 feet. At what altitude was the plane flying?

7. Some number added to $^{-}11$ is 37. Divide this number by $^{-}12$. Then, multiply by $^{-}8$. What is the final number?

8. Jim decided to go for a drive in his car. He started out at 0 miles per hour (mph). He then accelerated 20 mph down his street. Then, to get on the highway he accelerated another 35 miles per hour. A car was going slow in front of him so he slowed down 11 mph. He then got off the highway, so he slowed down another 7 mph. At what speed is he driving?

© Carson-Dellosa • CD-704384

Name________________________________

7.NS.A.1d, 7.NS.A.3

Adding and Subtracting Rational Numbers

$$^{-}3 + {^{-}2} + 2\frac{1}{2} = {^{-}5} + 2\frac{1}{2} = {^{-}4}\frac{2}{2} + 2\frac{1}{2} = {^{-}2}\frac{1}{2}$$

1. $^{-}1.6 + 1\frac{7}{10} =$

   (Hint: $1\frac{7}{10} = 1.7$)

2. $0 - 6\frac{1}{2} + {^{-}3} =$

3. $\frac{^{-}3}{4} + 5 - \frac{1}{2} =$

4. $9 - 10.2 + {^{-}8.6} =$

5. $\frac{1}{2} + 1\frac{1}{2} - 1\frac{1}{3} =$

6. $6.75 - 3\frac{1}{2} + 2.55 =$

   (**Hint:** $3\frac{5}{10} = 3.5$)

7. $3\frac{3}{7} - {^{-}1}\frac{1}{7} + \frac{3}{7} =$

8. $^{-}7 - {^{-}2}\frac{3}{4} + {^{-}5}\frac{1}{4} =$

9. $7\frac{1}{10} + {^{-}7.25} - 11.39 =$

10. $^{-}8\frac{1}{4} + {^{-}3}\frac{3}{12} - 7\frac{2}{3} =$

11. $^{-}5 - 7\frac{1}{8} + {^{-}3}\frac{5}{12} =$

12. $3\frac{3}{10} + {^{-}3.38} - 6\frac{6}{10} =$

© Carson-Dellosa • CD-704384

Name________________________

7.NS.A.1d, 7.NS.A.3

More Adding and Subtracting Rational Numbers

1. $^{-}3\frac{5}{10} + 8 =$

2. $^{-}5\frac{3}{7} + {}^{-}3\frac{3}{14} =$

3. $6\frac{1}{6} - 6\frac{3}{10} =$

4. $^{-}8 + 15.32 =$

5. $^{-}8\frac{3}{10} - {}^{-}5.9 =$

6. $13 - 5\frac{3}{5} =$

7. $12\frac{1}{9} + {}^{-}5\frac{2}{3} =$

8. $^{-}11.03 - {}^{-}21.6 =$

9. $^{-}7\frac{3}{10} - 16.53 =$

10. $31\frac{8}{9} + {}^{-}27\frac{27}{81} =$

11. $11 - 18.6 + {}^{-}3\frac{3}{10} =$

12. $^{-}5\frac{2}{10} + 16.7 - 3\frac{1}{5} =$

13. $13\frac{1}{3} + {}^{-}12 + {}^{-}7\frac{7}{12} =$

14. $41.32 + {}^{-}18.7 - 16.21 =$

15. $^{-}18.75 - 5\frac{3}{4} - 7\frac{5}{12} =$

16. $^{-}15 - 21\frac{1}{7} + 18\frac{2}{49} =$

17. $7\frac{2}{3} + {}^{-}8\frac{4}{9} - {}^{-}16\frac{1}{6} =$

18. $^{-}31.5 - {}^{-}3\frac{7}{10} + 21 =$

19. $25\frac{1}{5} - 17.3 + {}^{-}11\frac{2}{11} =$

20. $19.25 - {}^{-}6\frac{3}{4} + 12\frac{5}{12} =$

© Carson-Dellosa • CD-704384

Name________________________________ 7.NS.A.2a, 7.NS.A.2b, 7.NS.A.2c, 7.NS.A.3

Multiplying and Dividing Rational Numbers

$$^-4 \cdot 5 \cdot \frac{1}{2} = {}^-20 \cdot \frac{1}{2} = -\frac{\cancel{20}^{\,10}}{1} \cdot \frac{1}{\cancel{2}_{\,1}} = -\frac{10}{1} = {}^-10$$

$$5\frac{1}{4} \cdot 1\frac{2}{7} \div 1\frac{1}{2} = \frac{21}{4} \cdot \frac{9}{7} \div \frac{3}{2} = \frac{\cancel{21}^{\,3}}{\cancel{4}_{\,2}} \cdot \frac{\cancel{9}^{\,3}}{\cancel{7}_{\,1}} \cdot \frac{\cancel{2}^{\,1}}{\cancel{3}_{\,1}} = \frac{9}{2} \text{ or } 4\frac{1}{2}$$

1. $^-1\frac{2}{3} \cdot {}^-3\frac{1}{5} =$

2. $4\frac{5}{9} \div -\frac{10}{27} =$

3. $4\frac{1}{4} \cdot 3\frac{1}{5} =$

4. $^-9\frac{3}{8} \div {}^-3\frac{9}{12} =$

5. $-\frac{3}{8} \cdot 4 \cdot \frac{4}{9} =$

6. $^-9\frac{3}{5} \div \frac{12}{5} \cdot {}^-4 =$

7. $^-4.1 \cdot {}^-5.2 \div 4 =$

8. $6.2 \cdot 3 \cdot -\frac{1}{2} =$

9. $({}^-2\frac{1}{2})\,({}^-2\frac{1}{2}) \div 0.5 =$

10. $-\frac{6}{7} \cdot -\frac{5}{12} \cdot -\frac{2}{15} =$

11. $5\frac{2}{3} \cdot 9.81 \cdot 0 =$

12. $12 \cdot 3\frac{1}{4} \cdot {}^-2\frac{2}{3} =$

Name________________________________

7.NS.A.2a, 7.NS.A.2b, 7.NS.A.2c, 7.NS.A.3

Working with Rational Numbers

1. $^{-}9\frac{3}{5} \cdot \frac{5}{12} =$

2. $-\frac{16}{7} \div \frac{12}{35} =$

3. $4\frac{1}{2} \cdot {}^{-}2\frac{2}{7} =$

4. $^{-}5\frac{5}{6} \div 2\frac{1}{3} =$

5. $^{-}8\frac{1}{3} \cdot {}^{-}2\frac{2}{5} =$

6. $16\frac{1}{8} \div 14\frac{1}{3} =$

7. $^{-}37.6 \cdot 0.03 =$

8. $^{-}16.188 \div {}^{-}4.26 =$

9. $^{-}1.75 \cdot {}^{-}3.4 =$

10. $^{-}3.45 \div 1\frac{1}{2} =$

11. $^{-}8 \div {}^{-}1\frac{1}{3} \cdot {}^{-}5 =$

12. $4.498 \div {}^{-}1.73 \cdot {}^{-}1.2 =$

13. $-\frac{5}{7} \div -\frac{1}{14} \cdot -\frac{1}{2} =$

14. $^{-}6\frac{2}{3} \cdot 2.75 \div {}^{-}1\frac{2}{3} =$

15. $-\frac{3}{8} \div {}^{-}3 \cdot \frac{4}{5} =$

16. $12\frac{3}{8} \cdot {}^{-}2\frac{2}{3} \div 2.5 =$

17. $-\frac{5}{6} \cdot 4\frac{1}{4} \cdot -\frac{3}{5} =$

18. $^{-}3\frac{1}{5} \div 4\frac{2}{5} \div {}^{-}1\frac{1}{7} =$

19. $3\frac{3}{5} \cdot {}^{-}1.46 =$

20. $4\frac{2}{3} \div -\frac{6}{7} \cdot \frac{9}{10} =$

© Carson-Dellosa • CD-704384

Name________________________________ 7.NS.A.2a, 7.NS.A.2b, 7.NS.A.2c, 7.NS.A.3

Order of Operations with Rational Numbers

Order of operations: Perform operations within parentheses and brackets.
Compute exponents.
Multiply or divide in order from left to right.
Add or subtract in order from left to right.

$$2 + {}^{-}3 \cdot 5 = 2 + {}^{-}15 = {}^{-}13$$

$$7 - 6^2 \div 2 \cdot 5 = 7 - 36 \div 2 \cdot 5 = 7 - 18 \cdot 5 = 7 - 90 = {}^{-}63$$

1. ${}^{-}28 \div 7 + 2\frac{1}{3} =$

2. $\frac{1}{2}({}^{-}16 - 4) =$

3. ${}^{-}9 \div {}^{-}3 + 4 \cdot -\frac{1}{4} - 20 \div 5 =$

4. $\frac{1}{3}[({}^{-}18 + 3) + (5 + 7) \div {}^{-}4] =$

5. $(8\frac{1}{3} + 3\frac{2}{3}) \div 4 - {}^{-}16 =$

6. $\dfrac{(80 \cdot \frac{1}{2}) + 35}{{}^{-}10 + 25} =$

7. $2\,[{}^{-}6\,(3 - 12) - 17] =$

8. $\frac{1}{4}(20 + 72 \div {}^{-}9) =$

9. $3 \cdot 2\,[4 + (9 \div 3)] =$

10. $50 \div [(4 \cdot 5) - (36 \div 2)] + {}^{-}91 =$

Name____________________________________ 6.EE.A.1

Calculator Order

Use a scientific calculator to solve each problem. Turn the calculator around to determine the word answer.

Problem	Solution	Clue	Word
1. 501 x 7		To not win	
2. $10^3 - 3 \times 131$		Type of cabin	
3. $17^2 + 7^2$		It buzzes.	
4. 67,077 ÷ 87		Sick	
5. 2 • (2 • 1900 + 3 • 23)		It rings.	
6. $2^9 + 2$		Not hers	
7. $279^2 - (500 - 4)$		Nautilus ________	
8. $3^3 \times 100 + 3 \times 115$		Worn on foot	
9. $22{,}416 \div 2^2$		Big pigs	
10. 473,720 – 12,345		Snow vehicle	
11. 3 x 5 x 246 +15		Bottom of shoe	
12. 4,738 – 1,234		Fire equipment	
13. $60^2 + 4 \times 26$		Center of a donut	
14. 11 x (60 – 2)		To plead	
15. 5787 ÷ 9 x 12		Fish organ	
16. 12,345 + 23,456 – 465		They "honk."	
17. 8 x 100 + 8 – 1		Tennis shot	
18. $50 \times 700 + 3 \times 6^2$		Capital of Idaho	
19. 50 x 110 + (10 – 3)		Not a win	
20. $64{,}118 - 80^2$		Ducks' beaks	

A googol is 10^{100} or 1 followed by 100 zeros.
What number would result in the "calculator word" googol?________________

© Carson-Dellosa • CD-704384

Name________________________________ 6.NS.B.3, 6.NS.C.7a, 6.NS.C.7b

Comparing Rational Numbers

Use <, >, or = to make each a true sentence.

5.68	____	5.7	$^{-}7\frac{3}{10}$	____	$^{-}7.29$
5.68	<	5.70	$^{-}7.30$	<	$^{-}7.29$

1. 2.5 ______ $2\frac{17}{34}$
2. 1.049 ______ 1.49
3. $^{-}0.\overline{3}$ ______ $^{-}0.3$
4. 15.62 ______ 1.562
5. 8156.6 ______ 8166.6
6. $^{-}7\frac{4}{5}$ ______ $^{-}7\frac{24}{30}$
7. $^{-}8\frac{7}{8}$ ______ $^{-}8.857$
8. 329.93 ______ 32.993
9. 982.61 ______ 7662.8
10. $13\frac{5}{8}$ ______ 13.6

$5\frac{1}{2}, 5\frac{3}{5}, 5.4$ → Rewrite → 5.5, 5.6, 5.4

Descending Order: $5\frac{3}{5}, 5\frac{1}{2}, 5.4$

1. 6.41, 6.411, 6.4111
2. $^{-}2\frac{9}{14}, ^{-}2\frac{5}{8}, ^{-}2\frac{4}{7}$
3. 11.6, $11\frac{2}{3}, 11\frac{14}{25}$
4. $^{-}0.030, -\frac{33}{100}, ^{-}0.003$
5. $7\frac{5}{8}, 7\frac{3}{4}, 7.775$
6. $^{-}10\frac{3}{4}, ^{-}10.82, ^{-}10\frac{2}{3}$
7. 3.08, $3\frac{4}{5}, 3\frac{3}{5}$
8. $^{-}1.35, ^{-}1\frac{1}{8}, ^{-}1\frac{1}{4}$

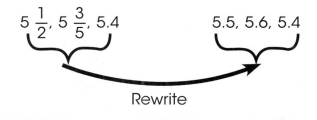

Name________________________________ 6.NS.C.7c, 6.NS.C.7d, 7.NS.A.1c, 7.NS.A.3

Opposites and Absolute Values

$^-(5c + 9d) = {}^-5c - 9d$ $\quad$ $^-|7 - 9| = {}^-|{}^-2| = {}^-2$

1. $|{}^-12| =$

2. $-|5\frac{1}{2}| =$

3. $|{}^-5| + |9| =$

4. $7 + |{}^-3| =$

5. $|7| + |{}^-7| =$

6. $-({}^-3 + 4) =$

7. $-(9 - 9) =$

8. $|9| - |{}^-12| =$

9. $|{}^-3| + |9| - 6 =$

10. $^-3|5| - |5| =$

11. $|{}^-25| - |{}^-14| =$

12. $^-18 + [-({}^-13)] =$

13. $|1 - 3| + 5 =$

14. $\dfrac{-|{}^-3 + 5|}{{}^-9 + [{}^-({}^-1)]} =$

15. $-[2n - ({}^-7)] =$

16. $-({}^-2x + {}^-3y) =$

17. $-(6x - 4y) =$

18. $10m - ({}^-2n) =$

19. $^-2(3m^2 - 2m - 1) =$

20. $^-3(4x - 6y) =$

© Carson-Dellosa • CD-704384

Name________________________________ 7.NS.A.1d, 7.NS.A.2a, 7.NS.A.2c, 7.NS.A.3

Test Prep: Real Numbers

1. 1 – 3 equals

   a. $^-4$ b. $^-2$ c. 4 d. 2

2. If x and y are positive integers and if $\frac{x}{y} = 1$ and $(x + y)^2 = z$, which of the following can equal z?

   a. 5 b. 9 c. 16 d. 25

3. $(^-1)\ (^-2)\ (^-3)\ (+4) =$

   a. $^-10$ b. 24 c. $^-24$ d. $^-36$

4. $(^-2) - (^-5) =$

   a. $^-7$ b. $^-3$ c. 3 d. 7

5. $(^-5) + (^-2) =$

   a. $^-7$ b. $^-3$ c. 3 d. 7

6. $(\frac{1}{2}) \div (-\frac{7}{8}) =$

   a. $-\frac{4}{7}$ b. $-\frac{7}{16}$ c. $^-1\frac{3}{4}$ d. $^-2\frac{2}{7}$

7. $7 - [(^-8) + (^-2)]$

   a. $^-3$ b. 4 c. 13 d. 17

8. $\frac{(^-18) + (^-2)}{(7) + (^-2)}$

   a. $2\frac{2}{9}$ b. 4 c. $3\frac{1}{5}$ d. $^-4$

9. The integers $^-2$, $^-7$, 5, and $^-5$ written from least to greatest are

   a. $^-2, ^-5, ^-7, 5$ b. $^-5, ^-7, ^-2, 5$ c. $^-7, ^-5, ^-2, 5$ d. $^-7, ^-2, ^-5, 5$

10. Which of the following conditions will make $x - y$ a negative number?

    a. $y > x$ b. $x > y$ c. $y > 0$ d. $x = y$

Name________________________________

6.NS.C.6b, 6.NS.C.6c, 6.NS.C.8

Plotting Points

Connect each of the following ordered points.

(*x*, *y*) = (0, $^-1$)

- vertical move ⟹ down one
- horizontal move ⟹ no move

"Ancient History"

Start at (0, $^-1$)

(1, $^-1$)
(1, $^-3$)
(3, $^-3$)
(3, $^-1$)
(5, 0)
(8, 0)
(7, 1)
(9, 0)
(8, 2)
(5, 1)
(4, 3)
(3, 4)
(2, 3)
(1, 4)
(0, 3)
($^-1$, 4)
($^-2$, 3)
($^-3$, 4)
($^-4$, 3)
($^-5$, 1)
($^-8$, 2)
($^-5$, 0)
($^-3$, $^-1$)
($^-3$, $^-3$)
($^-1$, $^-3$)
($^-1$, $^-1$)
(0, $^-1$)
End

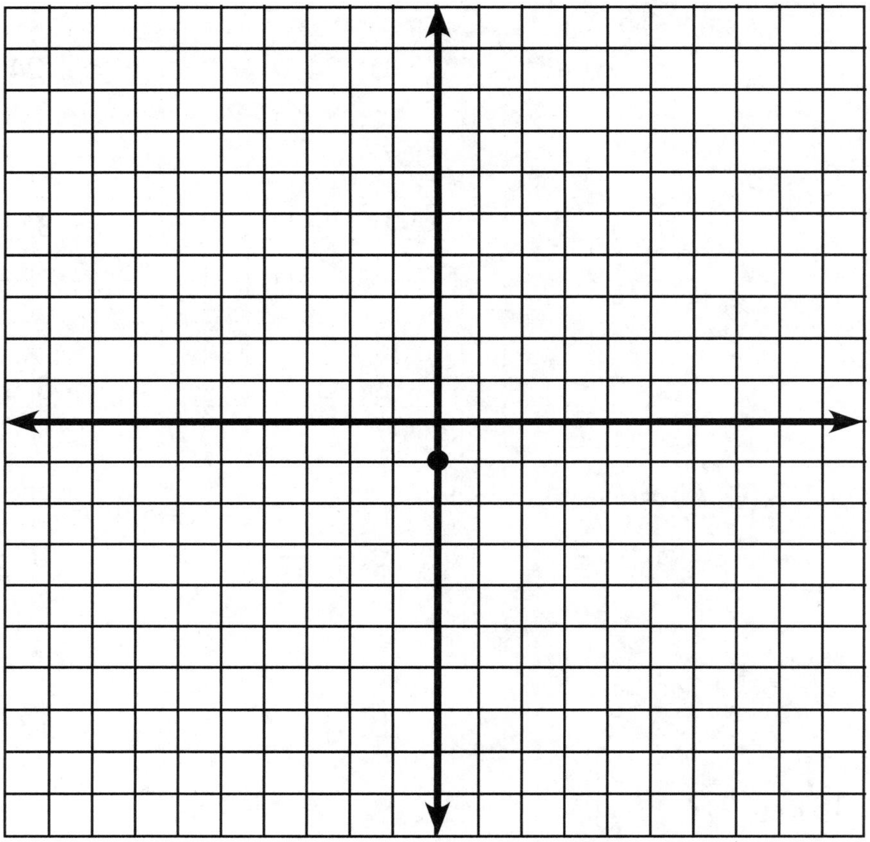

© Carson-Dellosa • CD-704384

Name________________________________

6.NS.C.6b, 6.NS.C.6c, 6.NS.C.8

Coordinates and Graphing

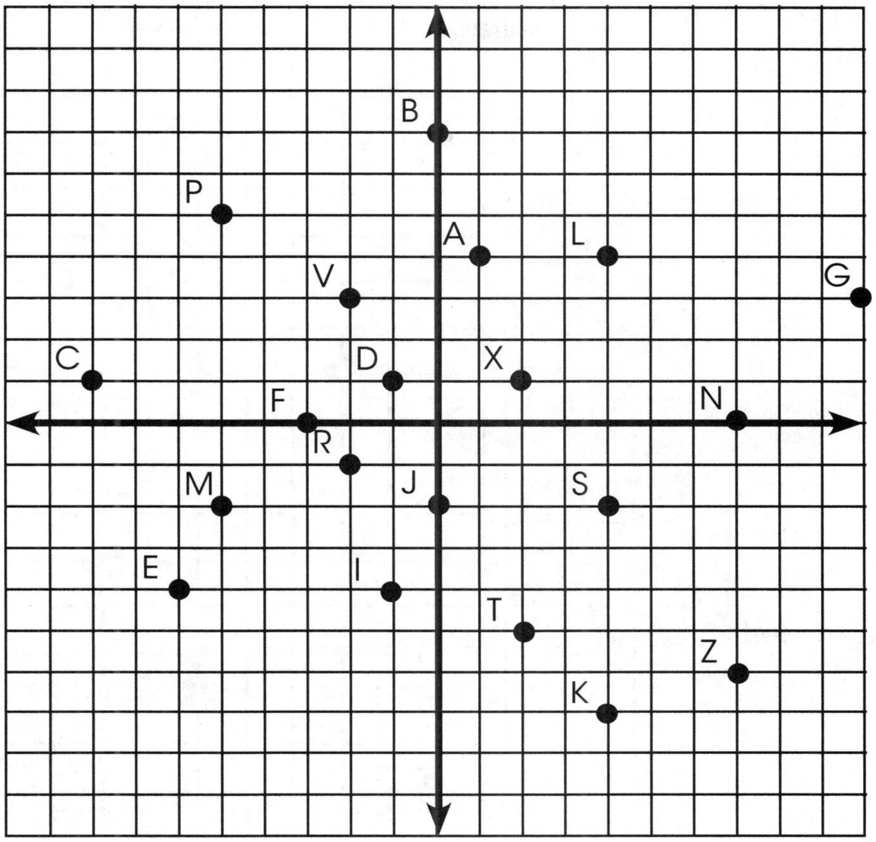

Find the coordinates associated with the following points.

1. A
2. K
3. E
4. P
5. T
6. C
7. B
8. S
9. D
10. N

Find the letter associated with each pair of coordinates.

11. (2, 1)
12. (-1, -4)
13. (10, 3)
14. (7, -6)
15. (-2, -1)
16. (-2, 3)
17. (-3, 0)
18. (4, 4)
19. (-5, -2)
20. (0, -2)

Name________________________________

6.NS.C.6b, 6.NS.C.6c, 6.NS.C.8

Graphing with Ordered Pairs

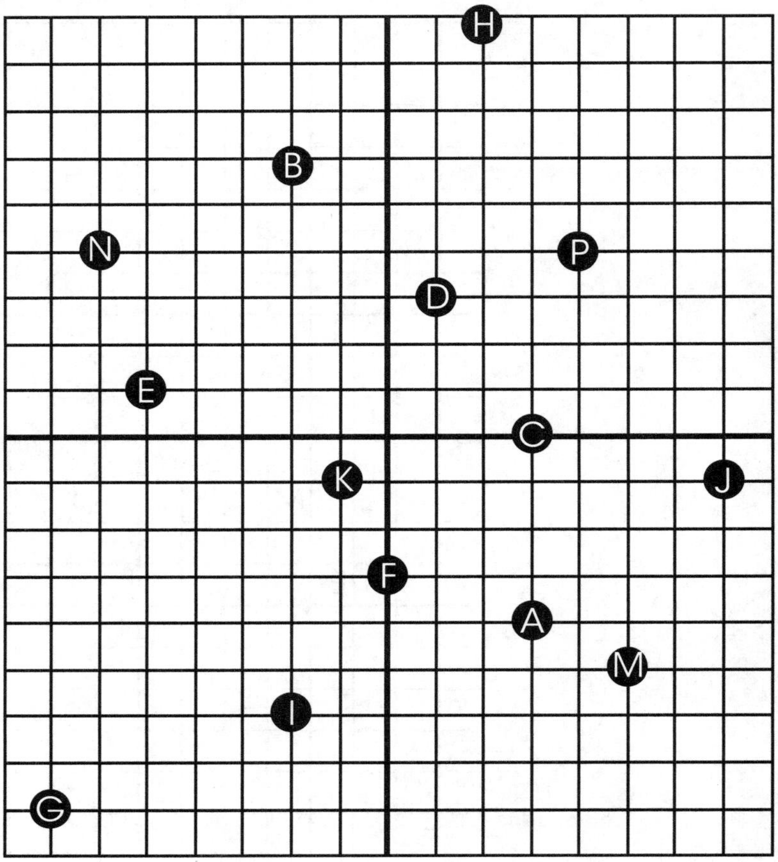

I. Find the coordinates of the indicated point.

1. A
2. I
3. H
4. C
5. E
6. N

II. Name the graph (letter) of each ordered pair.

7. ($^{-}2$, 6)
8. (0, $^{-}3$)
9. (5, $^{-}5$)
10. ($^{-}1$, $^{-}1$)
11. ($^{-}7$, $^{-}8$)
12. (7, $^{-}1$)
13. The coordinates are equal.
14. The *y*–coordinate is three times the *x*–coordinate.

III. Name the quadrant or axis on which each point lies.

15. ($^{-}4$, 3)
16. (0, 6)
17. (4, $^{-}2$)
18. ($^{-}1$, $^{-}1$)
19. ($^{-}2$, 0)
20. (1, 2)

© Carson-Dellosa • CD-704384

Name________________________________

Just for Fun

Mike, Dale, Paul, and Charlie are the athletic director, quarterback, pitcher, and goalie, but not necessarily in that order. From these four statements, identify the man in each position.

1. Mike and Dale were both at the ballpark when the rookie pitcher played his first game.

2. Both Paul and the athletic director had played on the same team in high school with the goalie.

3. The athletic director, who scouted Charlie, is planning to watch Mike during his next game.

4. Mike doesn't know Dale. One of these men is a quarterback.

	Quarterback	Goalie	Pitcher	Athletic Director
Mike				
Dale				
Paul				
Charlie				

Name________________________________ 6.EE.A.1

Exponents

I. Write in exponential form.

$4 \cdot x \cdot x \cdot y \cdot y \cdot y = 4x^2y^3$ The cube of $c - 4 = (c - 4)^3$

1. $a \cdot a \cdot a \cdot b$
2. $mn \cdot mn \cdot mn \cdot mn$
3. $9 \cdot x \cdot x \cdot x \cdot x \cdot x \cdot y \cdot y \cdot z$
4. $5\ (c + 1)\ (c + 1)\ (c + 1)$
5. $(a + b)$ squared
6. The quotient of 3 and the cube of $y + 2$
7. $x \cdot x \cdot y \cdot y \cdot y \cdot y \cdot z$
8. $(-x)\ (-x)\ (-x)$
9. $3 \cdot ab \cdot ab \cdot ab \cdot ab$
10. The square of $x^2y - 3$

II. Evaluate each expression if $x = {}^-1$, $y = 2$, $z = {}^-3$.

$5x^2z^2 = 5 \cdot x \cdot x \cdot z \cdot z = 5 \cdot {}^-1 \cdot {}^-1 \cdot {}^-3 \cdot {}^-3 = 45$

1. x^5
2. x^2yz
3. $4y^3z$
4. $x^5y^4z^3$
5. $-(xyz)$
6. $10z^5$
7. x^2yz^2
8. ${}^-2xy^2$
9. $\frac{x^2z^2}{z}$
10. $11x^2$

© Carson-Dellosa • CD-704384

Name________________________________

6.EE.A.3, 6.EE.A.4, 7.EE.A.1, 7.EE.A.2

Combining Like Terms

$8x + 5y + {}^{-}17x = -9x + 5y$

1. $9x + 4x =$
2. $17x + x =$
3. $m + ({}^{-}4m) =$
4. ${}^{-}7x - 8x =$
5. $14a - 19a =$
6. $-a + 9a =$
7. $6xy + 5xy =$
8. ${}^{-}9m - m =$
9. $15a + ({}^{-}11a) =$
10. ${}^{-}14x + 13x =$
11. $5x^2y + 13x^2y =$
12. $21xy + ({}^{-}9xy) =$
13. $17x + 1 =$
14. $3.5y - 7.2y =$
15. ${}^{-}4.7y - 2.3y =$
16. $3a + 5c - 9a =$
17. $2x - 9x + 7 =$
18. $7x - 8 - 11x =$
19. $3x - 3y - 9x + 7y =$
20. $17x + 4 - 3x =$
21. $3x - 7y - 12y =$
22. $11a - 13a + 15a =$
23. $17x + 5a - 3x - 4a =$
24. $6x + 9y + 2x - 8y + 5 =$
25. $3xy + 4xy + 5x^2y + 6xy^2 =$
26. ${}^{-}25y - 17y + 6xy - 3xy =$

Name____________________________________ 6.EE.A.3, 6.EE.A.4, 7.EE.A.1, 7.EE.A.2

What's Not to Like?

Simplify each expression by combining like terms. Circle the expression in each problem that does not belong. Place the letter above the problem number below.

1. A. $5t + 3r + 9t - 10r$ E. $r + t - 8r + 13t$ I. $-r + 4t + 10t + 8r$

2. D. $12x - 3y + x + 2y$ E. $3\ (4x - 3y) + x + 3y$ F. $4\ (4x - 2y) - 3x + 7y$

3. E. $4\ (y - 7x) - y$ I. ${}^{-}30x - ({}^{-}2x)$ O. ${}^{-}7\ (4x + y) + 7y$

4. U. $6\ (x - y) - 3\ (3x + y)$ V. $3\ (3x - y) - 6y$ W. $4x + y - 7x - 10y$

5. Q. $3\ (r - 1) - 4r + 5$ X. $2\ (3 - 2r) - 4\ (2 - r)$ Z. $-r + 7 + 3r - 9 - 2r$

6. L. $8\ (x + y) + 3\ (x + y)$ M. $10\ (x + y) + x + y$ N. $9\ (x + y) - 2\ (x + y)$

7. A. $3\ (2b - a) - (2a - b)$ B. $3\ (a + 2b) - (b + 2a)$ C. $2\ (a + 2b) - (a - b)$

8. I. ${}^{-}5\ (a - b) - 2\ (a - b) + 8\ (a - b)$ U. $6\ (a - b) - 4\ (a - b) + (a - b)$ O. $(a - b) - (a - b) + (a - b)$

9. R. $3\ (x - y) - 2\ (y - x)$ S. $2\ (x - y) - 3\ (y - x)$ T. $3\ (y - x) - 2\ (x - y)$

10. L. ${}^{-}4\ [x + 2\ (5xy - x)]$ M. ${}^{-}4\ [x + 5\ ({}^{-}3xy + x)]$ N. ${}^{-}2\ [3x + 3\ ({}^{-}10xy + 3x)]$

Two expressions in each problem are

____	____	____	____	____	____	____	____	____	____
2	5	8	1	4	7	10	3	6	9

© Carson-Dellosa • CD-704384

Name________________________________ 6.EE.A.2c, 7.EE.B.3

Open Sentences

State the solution for each sentence.

$\frac{1}{2} \cdot {}^{-}10 = x$

$\frac{1}{\cancel{2}_{1}} \cdot \frac{\cancel{{}^{-}10}^{\,{}^{-}5}}{1} = x$

${}^{-}5 = x$

$\frac{{}^{-}56}{{}^{-}7} - 4 = z$

$8 - 4 = z$

$4 = z$

1. $\frac{18 + {}^{-}6}{2} = a$

2. ${}^{-}3 \cdot 4 - 6 = c$

3. $4.5 - 6.2 = p$

4. $\frac{{}^{-}3}{8} \cdot {}^{-}4 - 1 = q$

5. $\frac{{}^{-}15 + {}^{-}27}{3} = x$

6. ${}^{-}8.1 \cdot 4.2 + 16 = g$

7. $\frac{1}{3} \cdot {}^{-}15 + {}^{-}10 = r$

8. $1\frac{3}{5} \div \frac{16}{45} = d$

9. $5 \cdot 7.32 - 18.19 = n$

10. $\frac{3}{4} \cdot {}^{-}16 + 8.12 = z$

11. $\frac{{}^{-}40 + 15}{5} + 6 = b$

12. $-\frac{2}{5} \div \frac{4}{15} + {}^{-}2\frac{1}{2} = t$

Name________________________

6.EE.A.2c, 6.EE.B.5

More Open Sentences

Using the given value, state whether each problem is true or false.

$28 = r \cdot \frac{1}{4}$, if $r = {}^-108$

$28 \stackrel{?}{=} {}^-108 \cdot \frac{1}{4}$

$28 \stackrel{?}{=} -27 \Longrightarrow$ False

1. $7 + x = 3\frac{1}{2}$, if $x = {}^-3\frac{1}{2}$
2. $y + 15 \div 6 = {}^-1\frac{1}{2}$, if $y = {}^-3$
3. $\frac{f}{13} + {}^-3 = 0$, if $f = 69$
4. $2x - 5.45 = 0.97$, if $x = 3.21$
5. $8\frac{1}{3} + a = 15\frac{8}{15}$, if $a = 7\frac{2}{5}$
6. $8 + (z - 32) = {}^-10$, if $z = 16$
7. $11.5 + c = 28\frac{1}{4}$, if $c = 16\frac{3}{4}$
8. $y(5 + 11) + 8 = 41$, if $y = 2$
9. $3g + 5.26 - 11.9 = 12.64$, if $g = {}^-3$
10. $5 + -\frac{16}{k} = {}^-3$, if $k = 2$
11. $7\frac{1}{9} \div w = \frac{1}{18}$, if $w = 2\frac{17}{32}$
12. $\frac{3(2q - q)}{8} + 29 = 32$, if $q = 8$
13. $\frac{16.8 - 91.6}{m}$ 37.4, if $m = 2$
14. $11\frac{1}{4} - f = 5\frac{1}{16}$, if $f = 16\frac{5}{16}$

© Carson-Dellosa • CD-704384

Name________________________________ 6.EE.A.2c

Evaluating Expressions

Evaluate the following, if $a = \frac{1}{2}$, $x = 4$, and $y = {}^{-}2$.

$$5x\ (2a - 5y) = 5 \cdot 4\ (2 \cdot \frac{1}{2} - 5 \cdot {}^{-}2) = 20\ (1 + 10) = 20\ (11) = 220$$

1. $4\ (a - 1) =$

2. $4a - 3y =$

3. $4\ (x - 3y) =$

4. $x\ (a + 6) =$

5. $6a + {}^{-}12a =$

6. $7\ (x + -y) =$

7. $6a\ (8a + 4y) =$

8. $3x + 2\ (a - y) =$

9. $x\ (ax + ay) =$

10. $ay + y - 5ax =$

11. $xy\ (2a + 3x - 2) =$

12. $4x - (xy + 2) =$

13. $5y - 8a + 6xy - 7x =$

14. $10x\ (8a + {}^{-}4y) + {}^{-}3y =$

15. $6xy - 2x\ (4a - 8y) =$

16. $(2a - x)\ (2x - 6) =$

© Carson-Dellosa • CD-704384

Name____________________________

6.EE.A.3, 6.EE.A.4, 7.EE.A.1, 7.EE.A.2

Simplifying Expressions

Distributive Property

$$3\,(x + 2y) = 3x + 3 \cdot 2y$$
$$= 3x + 6y$$

1. $^{-}7\,(a + b) =$
2. $x\,(y - 4) =$
3. $-\frac{2}{3}\,(c - 12) =$
4. $^{-}8\,(\frac{t}{2} + 6) =$
5. $y\,(^{-}16 + 2x) =$
6. $3\,(2a - 8b) =$
7. $2x\,(3y + {}^{-}6) =$
8. $7\,(^{-}5x + 8z) =$
9. $^{-}5y\,(6z - 10) =$
10. $^{-}3x\,(^{-}7 + 8y) =$

Combining Like Terms

$$6m - 4m + 3p = (6 - 4)m + 3p$$
$$= 2m + 3p$$

same variable

1. $9y + 6y - 2 =$
2. $25x - x + 2y =$
3. $4a + 8b + 11a - 10b =$
4. $13xy + 18xy - 20xy =$
5. $^{-}2m + 16 - 13m =$
6. $4a + 7 + 3a - 8 - 3a =$
7. $16x + {}^{-}18y + 10x - 7y =$
8. $6c - 8ab + 9c - 10 =$
9. $18ab + {}^{-}6a + {}^{-}7b + 26ab + {}^{-}7b =$
10. $5x - 3x + 2xy + 31x + {}^{-}18xy =$

© Carson-Dellosa • CD-704384

Name________________________________ 6.EE.A.3, 6.EE.A.4, 7.EE.A.1, 7.EE.A.2

An Expression by Any Other Name

Simplify each expression. Cross out each box that contains an answer. The remaining words can be restated to make a familiar proverb.

1. $3\,(a + b) + 2b =$
2. $5a + 2a(5 - b) =$
3. $8 - 3(6 - 6a) =$
4. $4a + 6(a + 8) =$
5. $^{-}2a - 3(b - 4a) =$
6. $8(6a + 7b) - 11(2b + 8a) =$
7. $^{-}6(a + 5b) - 3(^{-}7b - a) =$
8. $2(a - b) + 3(a - b) - 4(a - b) =$
9. $4a + {}^{-}7(a + 2) =$
10. $6(a + 2b) + 8a - 16b =$
11. $3a + {}^{-}2(a + b) =$
12. $2(3a - 4b) - 6a =$
13. $^{-}5(2a - 3b) + 5(3b - 2a) =$
14. $4\,(11a - 9b) - 7(6a) =$
15. $^{-}3(4a - 5b) - (a - b) =$

$10 + 18a$ YOU	$^{-}10 + 18a$ ARE	$a - 2b$ SEE	$2a + 36b$ CANNOT	$3a + 5b$ LEAD	$12a - 8b$ INSTRUCT
$14a + 4b$ AN	$14a - 4b$ A	$40a + 34b$ ELDERLY	$^{-}40a + 34b$ HORSE	$3a - 14$ CANINE	$2a - 36b$ TO
$10a + 48$ WATER	$^{-}3a - 14$ BUT	$12a - 8b$ ON	$^{-}13a + 16b$ YOU	$^{-}3a - 9b$ CANNOT	$15a - 2b$ FRESH
$10a - 3b$ MAKE	$a - b$ HIM	$15a - 2ab$ DOWN	$^{-}20a + 30b$ DRINK	$20a - 30b$ PROCEDURES	$^{-}8b$ HOME

Write the familiar proverb.

_______ _______ _______ _______ _______ _______ _______ _______

Name________________________________

6.EE.A.3, 6.EE.A.4, 7.EE.A.1, 7.EE.A.2

Mixed Up Pairs

Solve each equation. Each equation in column A has the same solution as an equation in Column B. Find the pairs.

Column A		Column B
______	1. $y + 12 = 8$	A. $y - 12 = {}^-12$
______	2. $\frac{y}{6} = {}^-2$	B. $2y = {}^-8$
______	3. ${}^-7y = {}^-84$	C. $y - 1 = {}^-7$
______	4. ${}^-42 = y - 20$	D. $y - {}^-12 = 24$
______	5. $92 + y = 92$	E. $\frac{y}{{}^-4} = 3$
______	6. $9 = 54y$	F. $y + 2 = 11$
______	7. ${}^-12 = y - 6$	G. $y + 11 = {}^-11$
______	8. ${}^-1 = \frac{y}{20}$	H. $12y = 2$
______	9. $27 = 3y$	I. $\frac{y}{{}^-2} = {}^-10$
______	10. ${}^-5 + y = 15$	J. ${}^-15 = y + 5$

© Carson-Dellosa • CD-704384

Name________________________________

6.EE.A.3, 6.EE.A.4, 7.EE.B.3,

Solving Equations Using the Distributive Property

$$4\,(x - 3) = 20$$
$$4x - 12 = 20$$
$$4x - 12 + 12 = 20 + 12$$
$$\frac{4x}{4} = \frac{32}{4}$$
$$x = 8$$

1. $3(x + 8) = {}^{-}6$

2. $75 = {}^{-}5(a + 5)$

3. ${}^{-}8(y - 6) = {}^{-}16$

4. $20 = 4(\frac{t}{4} - 2)$

5. $17(x - 2) = {}^{-}34$

6. $63 = 9(2 - a)$

7. $6(2 - \frac{x}{6}) = 1$

8. ${}^{-}36 = 6(y - 2)$

9. ${}^{-}7(r + 8) = {}^{-}14$

10. $3(m + 5) = 42$

11. ${}^{-}54 = 3(2 + 5m)$

12. ${}^{-}3(x - 7) + 2 = 20$

Name________________________________ 6.EE.A.2a, 6.EE.B.6, 7.EE.A.2, 7.EE.B.4a

Writing Algebraic Expressions

The product of four and eleven	$4 \cdot 11$
A number increased by six	$x + 6$
The number divided by two	$y \div 2$ or $\frac{y}{2}$
Twice a number decreased by one	$2a - 1$

1. Five less than a number

2. Three times the sum of a number and twelve

3. Ten more than the quotient of c and three

4. Two increased by six times a number

5. Two–thirds of a number minus eleven

6. Twice the difference between c and four

7. The product of nine and a number, decreased by seven

8. Six times a number plus seven times the number

9. A number increased by twice the number

10. One–fourth times a number increased by eleven

11. The quotient of a number and three decreased by five

12. Twelve times the sum of a number and five times the number

© Carson-Dellosa • CD-704384

Name________________________________

6.EE.A.2a, 6.EE.B.6, 6.EE.B.7, 7.EE.B.3

Solving One-Step Problems

Write an equation and solve.

Nine more than a number is 33.
Find the number.

$$9 + n = 33$$
$$9 - 9 + n = 33 - 9$$
$$n = 24$$

1. A number decreased by 16 is $^-26$. Find the number.

2. One-fourth of a number is $^-60$. Find the number.

3. The product of negative eight and a number is 104. Find the number.

4. Twice a number is 346. Find the number.

5. A number increased by negative twenty-seven is 110. Find the number.

6. Tim weighs five pounds more than Mitchell. Find Mitchell's weight if Tim weighs ninety-three pounds.

7. The cost of five books is $71. What is the cost of each book?

8. The cost of a filter is $4. What is the cost of six filters?

Name____________________________________

6.NS.C.7a, 6.NS.C.7b, 6.EE.B.8

Graphing Inequalities

$x > 2$ 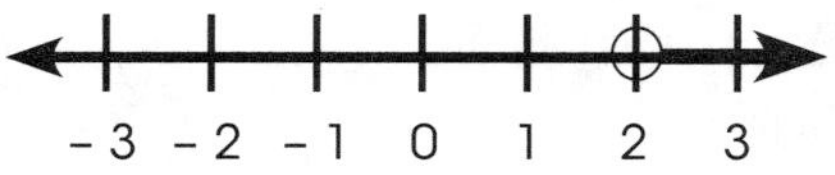

$y \leq 2$ 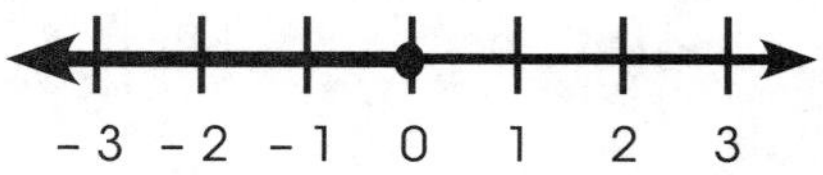

1. $x > 1$ 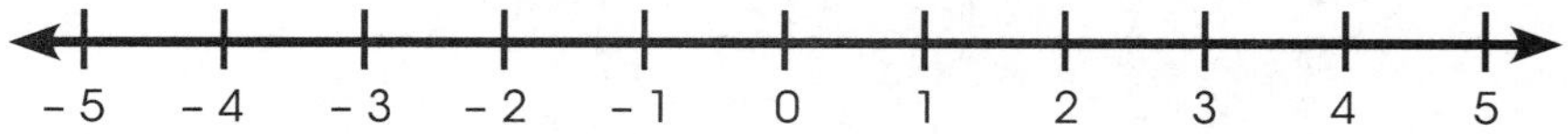

2. $a < {}^{-}1$ 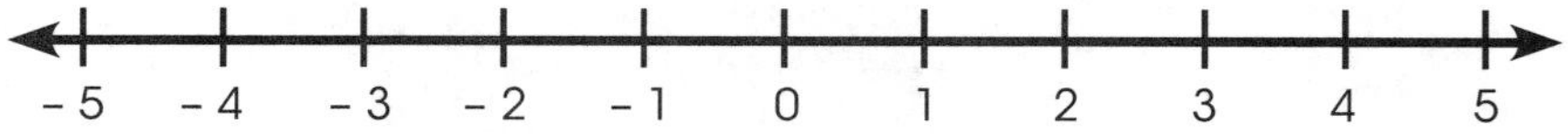

3. $y \leq 2$ 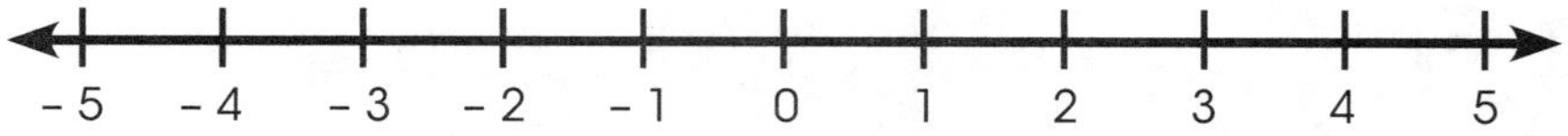

4. $b > {}^{-}4$ 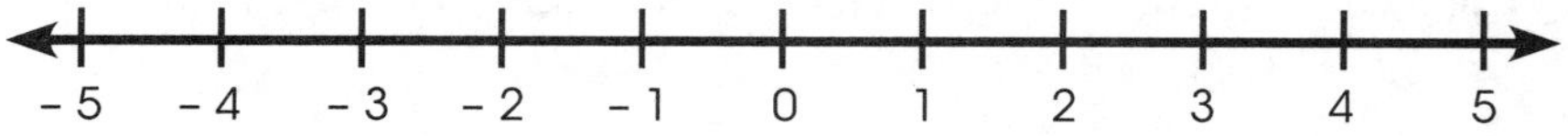

5. $p \geq 3$ 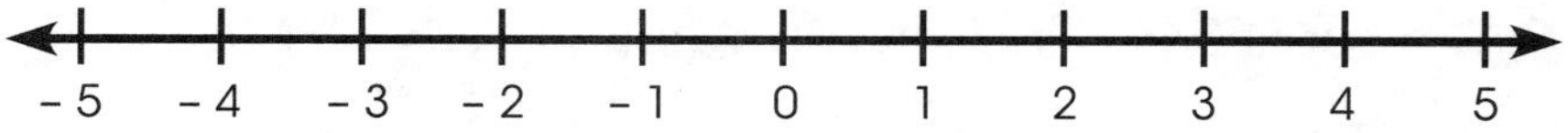

6. $x < \frac{1}{2}$ 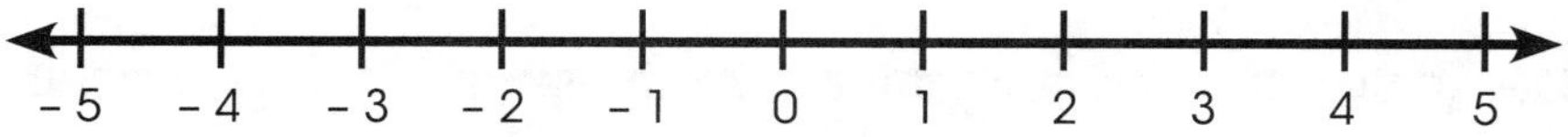

7. $y > {}^{-}1.5$ 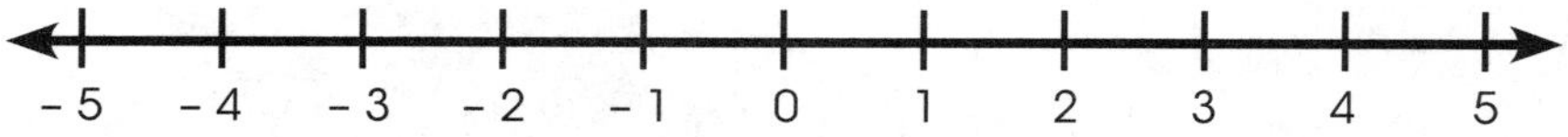

8. $m \leq 4\frac{1}{2}$ 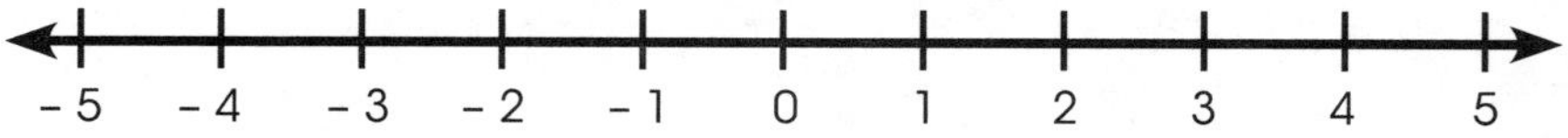

9. $c \leq \frac{{}^{-}10}{5}$ 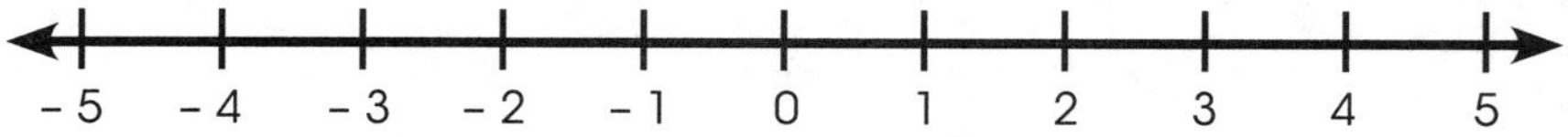

10. $d \geq 3.75$ 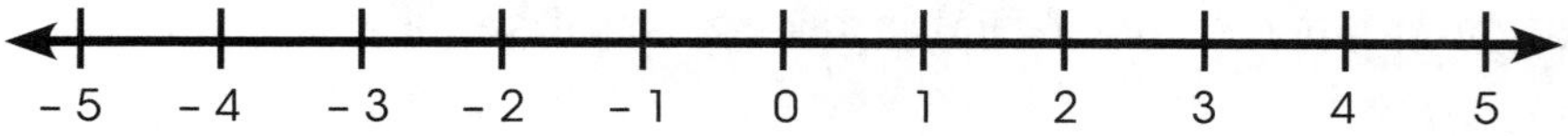

© Carson-Dellosa • CD-704384

Name________________________________ 7.EE.B.4b

Solving Inequalities with Addition or Subtraction

$m + 9 > 5$

$m + 9 - 9 > 5 - 9$

$m > {}^{-}4$

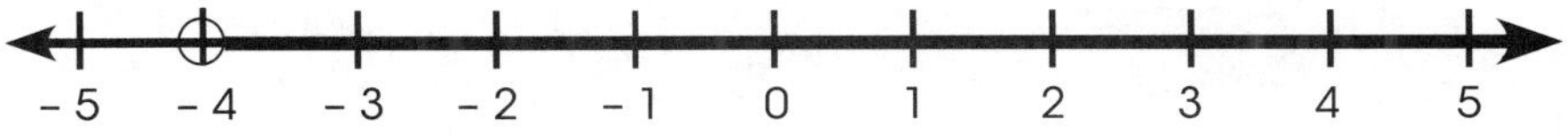

1. $g + 8 > 6$

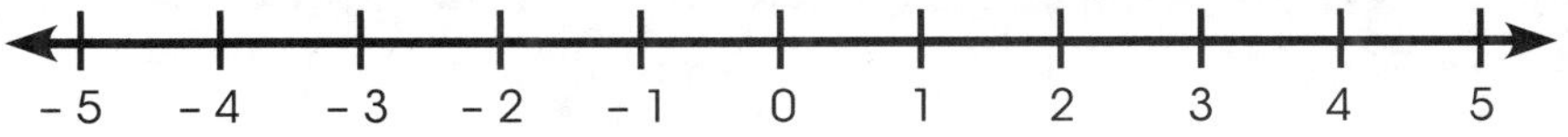

2. $d - 7 > {}^{-}3$

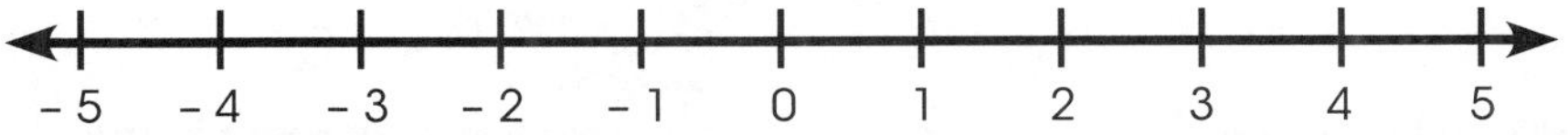

3. ${}^{-}3 > y + 1$

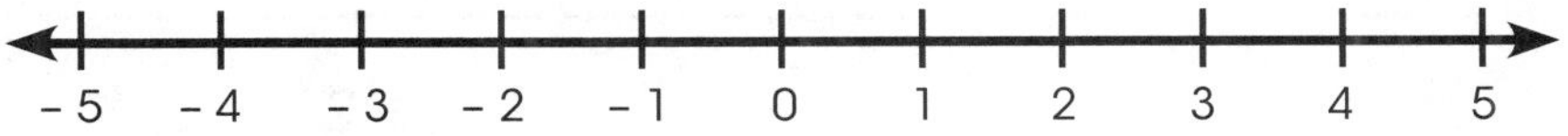

4. $a - 3 \leq 1$

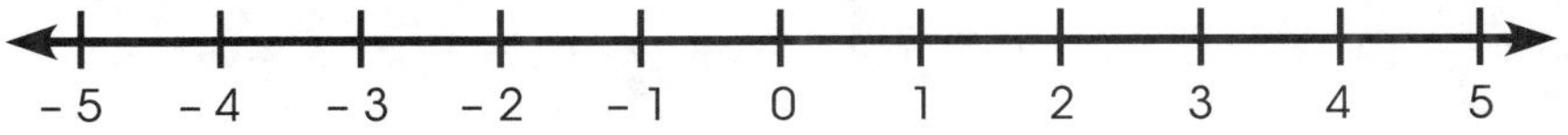

5. ${}^{-}4 \leq 1 + c$

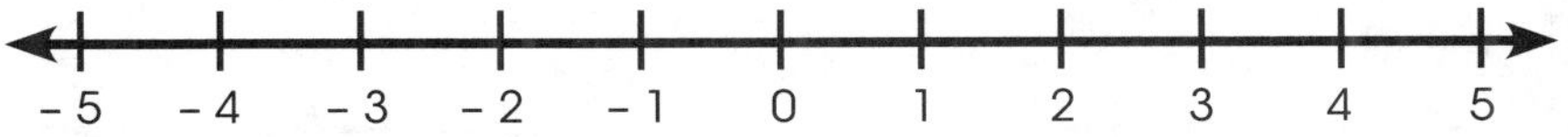

6. $x + \frac{1}{4} \geq 1\frac{1}{2}$

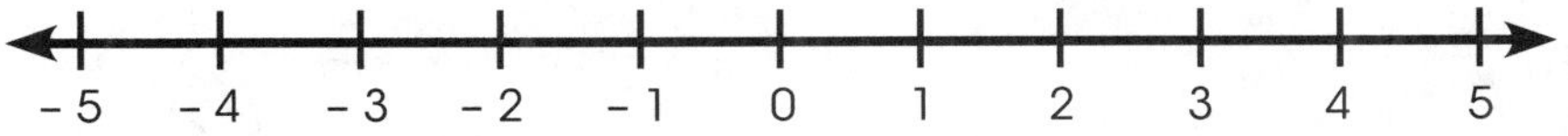

7. ${}^{-}2.4 < n - 0.6$

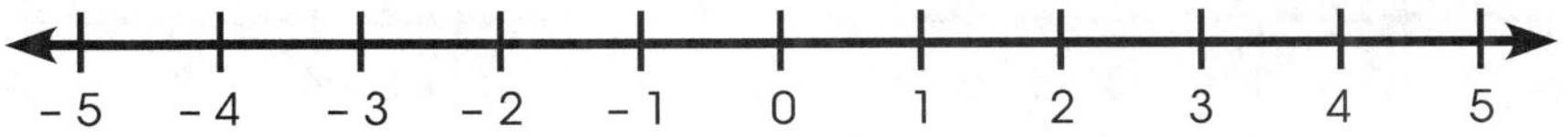

8. ${}^{-}20 + m \leq {}^{-}24$

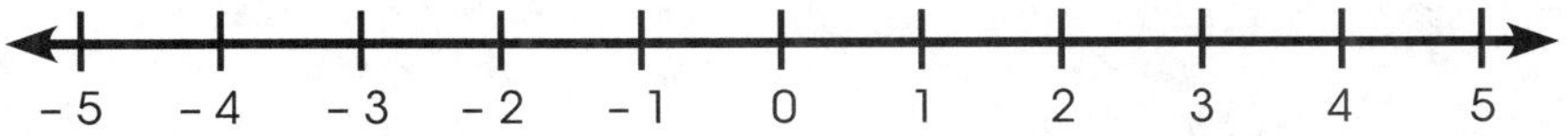

9. ${}^{-}7.5 + x \geq {}^{-}9$

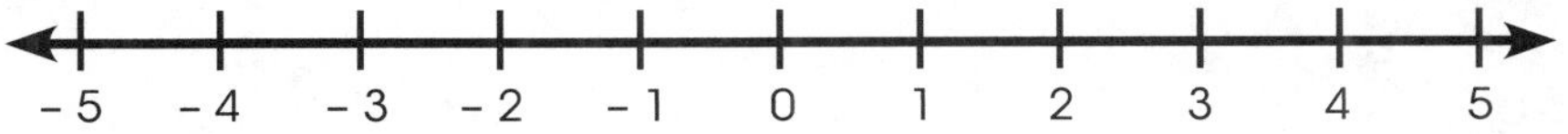

10. $3\frac{1}{3} \leq \frac{2}{9} + c$

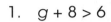

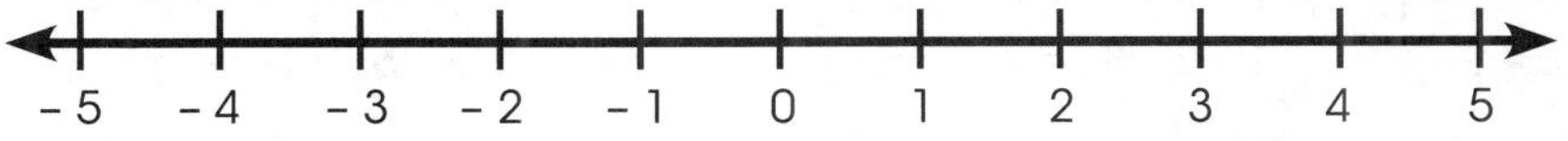

Name________________________________

7.EE.B.4b

Solving Inequalities with Multiplication or Division

$\frac{3y}{3} \leq \frac{9}{3}$

$y \leq 3$

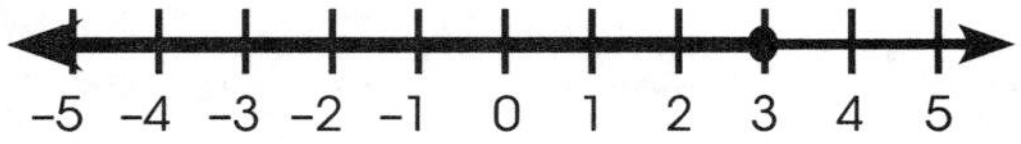

$\frac{40}{^-10} < \frac{^-10n}{^-10}$

$^-4 \circled{>} n$

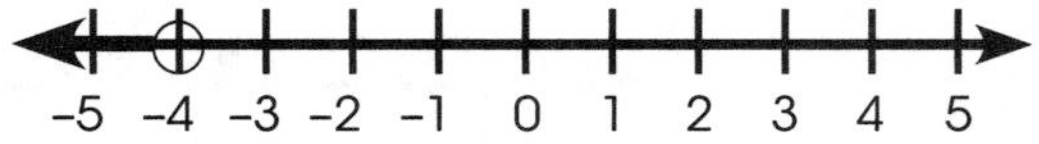

$-\frac{5}{2} \cdot -\frac{2}{5}x \geq {^-4} \cdot -\frac{5}{2}$

$x \circled{\leq} 10$

Note: division or multiplication by a negative number switches the sign.

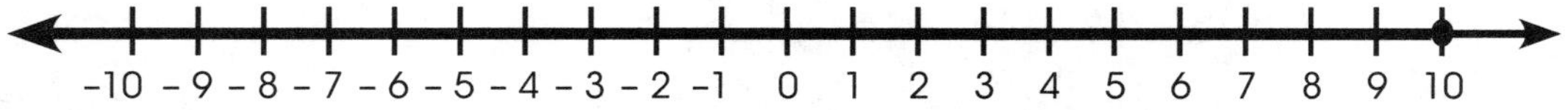

1. $11x > 22$

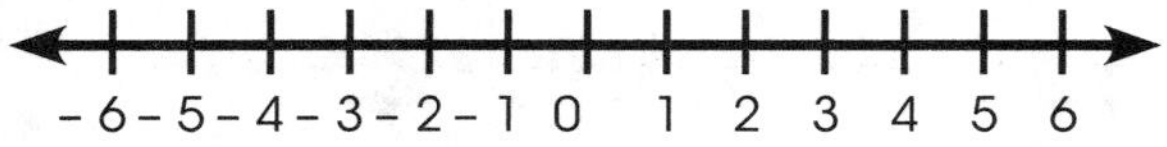

2. $^-15m \leq {^-75}$

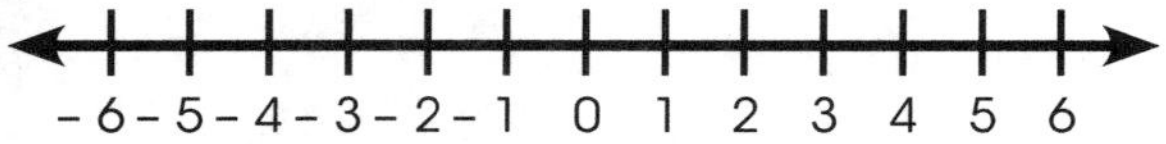

3. $^-1 > \frac{b}{3}$

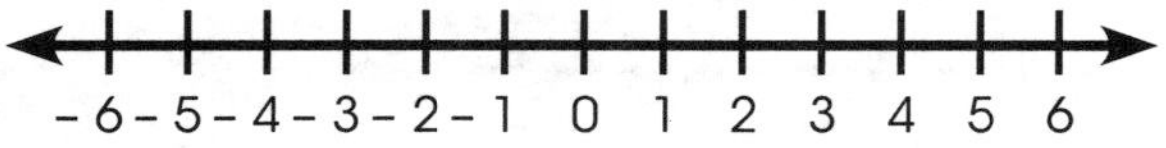

4. $1.9x \leq {^-7.6}$

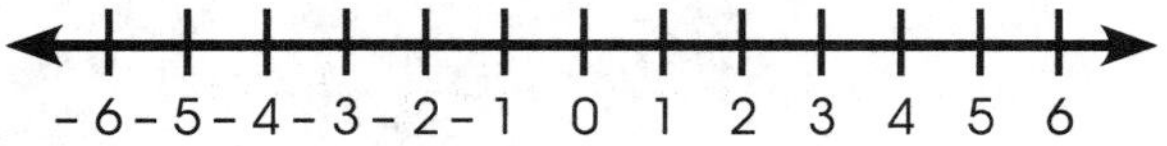

5. $\frac{3}{2}y < 6$

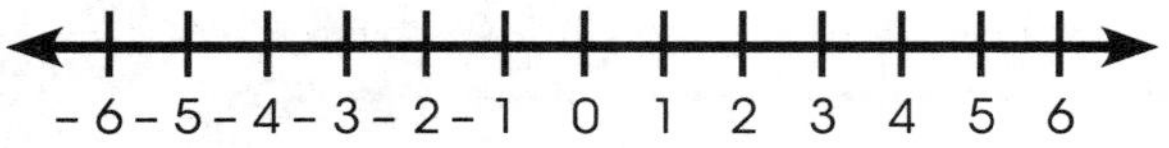

6. $^-26m \geq 13$

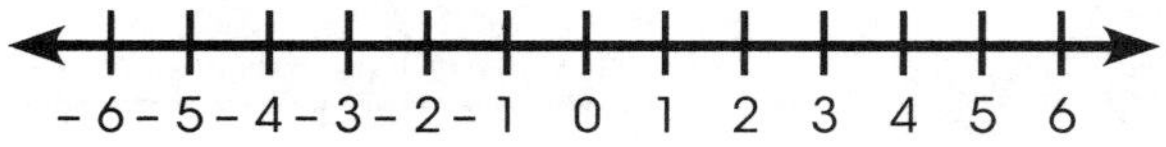

7. $^-4 \geq \frac{2}{3}x$

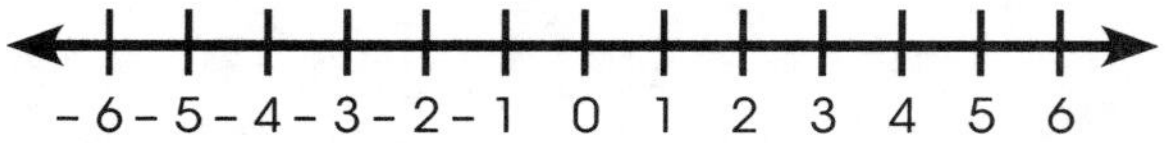

8. $^-2c < 2$

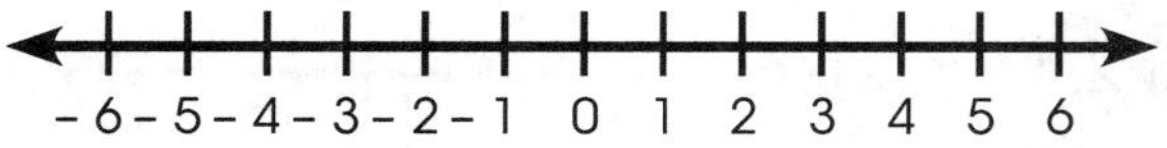

9. $^-3a \leq {^-9}$

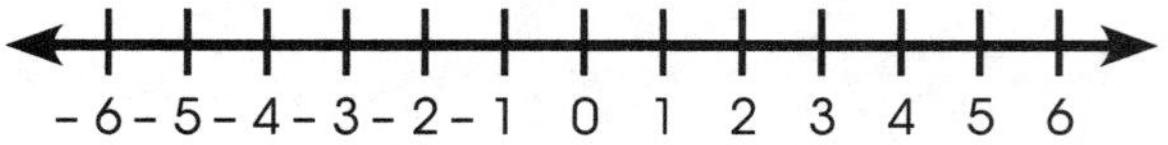

10. $-\frac{3}{4}x \geq {^-3}$

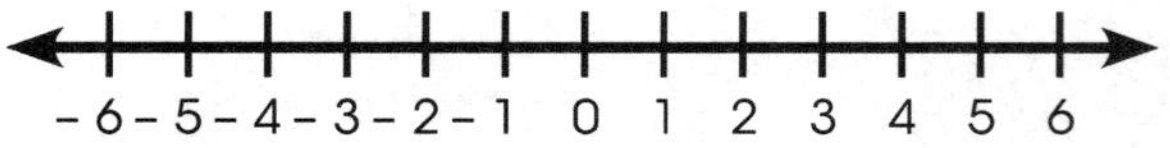

© Carson-Dellosa • CD-704384

Name________________________________ 7.EE.B.4b

Solving One-Step Inequalities

1. $a + 8 > 16$

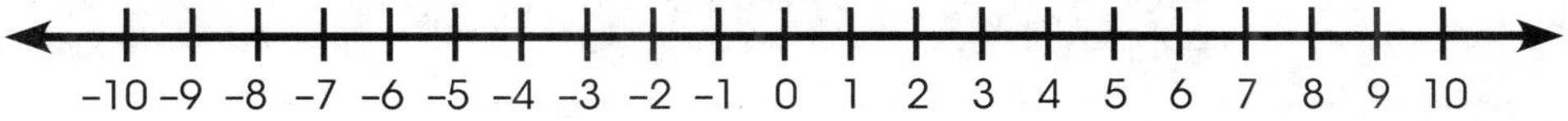

2. $^{-}7\frac{3}{5} \geq z - {}^{-}\frac{1}{15}$

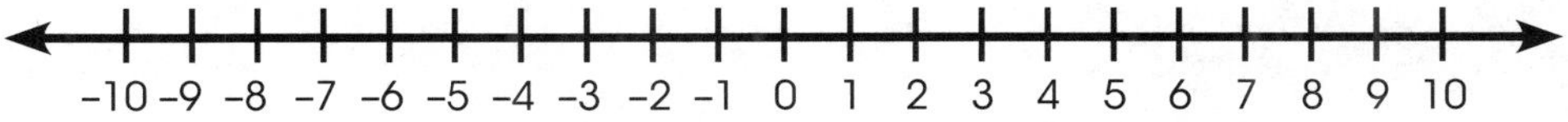

3. $^{-}28 < {}^{-}4x$

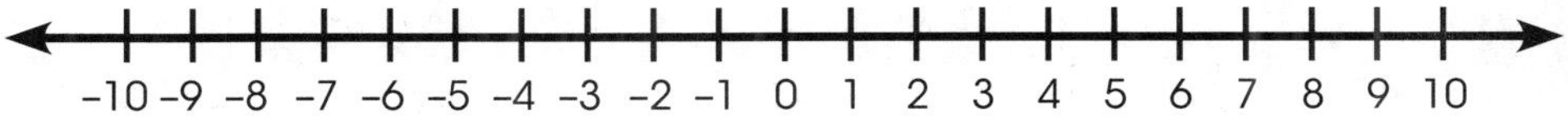

4. $^{-}28.5 \leq c + {}^{-}19.6$

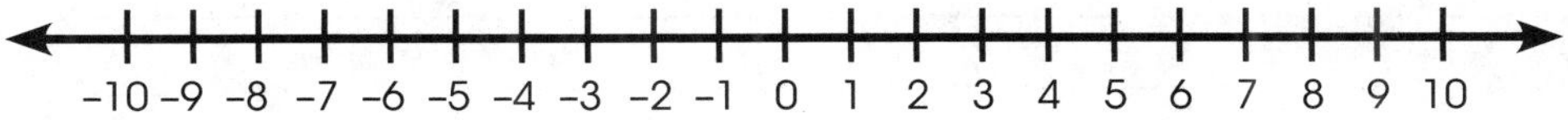

5. $\frac{^{-}4y}{3} > {}^{-}6$

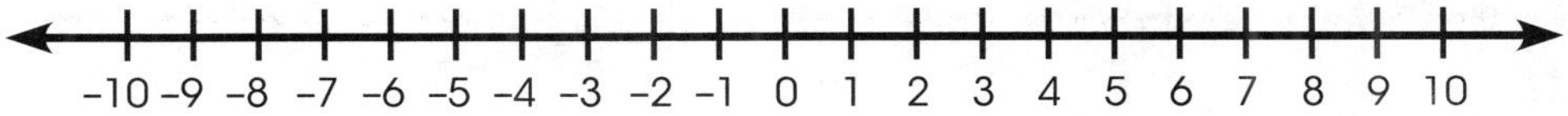

6. $6.3x < 7.56$

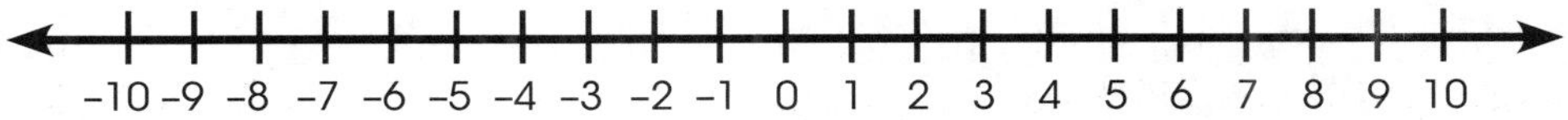

7. $-\frac{b}{3} \geq 3$

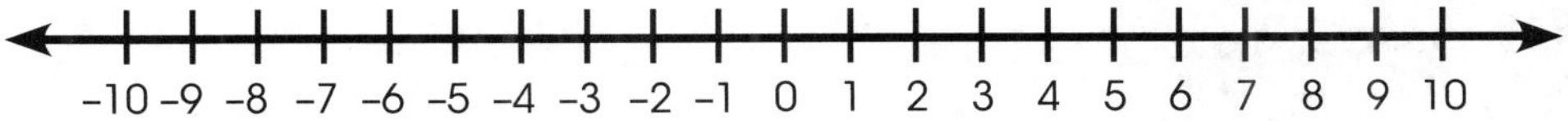

8. $^{-}18.2 < g - 13.7$

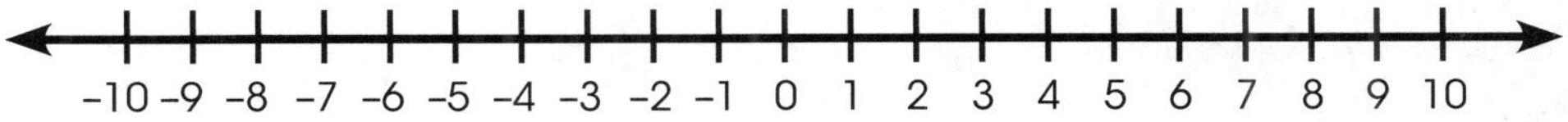

9. $b + 3\frac{1}{4} > {}^{-}2\frac{1}{8}$

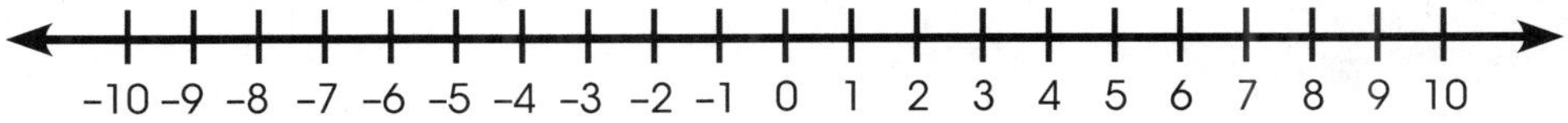

10. $\frac{z}{2} \leq {}^{-}3$

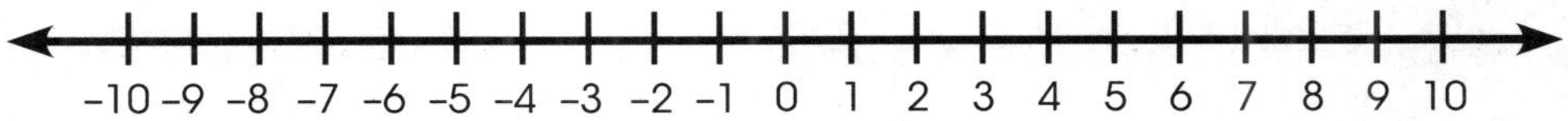

Name________________________________

7.EE.B.4b

Solving Inequalities with More than One Operation

$$^{-}14x + 8 \le 64$$

$$^{-}14x + 8 - 8 \le 64 - 8$$

$$\frac{^{-}14x}{^{-}14} + 8 \le \frac{56}{^{-}14}$$

$$x \ge {^{-}4}$$

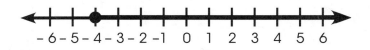

1. $7x - 1 < 20$

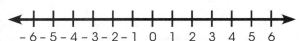

2. $^{-}4 + 2z \ge {^{-}8}$

3. $^{-}6x - 9 \ge {^{-}3}$

4. $4(2b - 3) \ge 36$

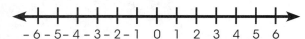

5. $7 < 5x - 8$

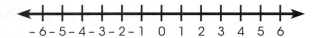

6. $^{-}17 > -7x - 45$

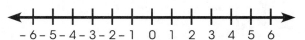

7. $^{-}5(2t - 1) \le 5$

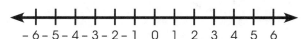

8. $8 - 4x > {^{-}12}$

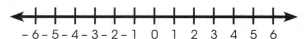

9. $^{-}2(2x - 1) \ge {^{-}9}$

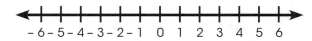

10. $41.56 < 6.3 - {^{-}8.2}x$

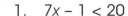

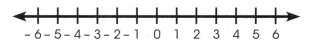

© Carson-Dellosa • CD-704384

Name____

7.EE.B.4b

Solving Inequalities with Variables on Both Sides

$$^{-}2a + 11 < a - 1$$

$$^{-}2a + 2a + 11 < a + 2a - 1$$

$$11 < 3a - 1$$

$$11 + 1 < 3a - 1 + 1$$

$$\frac{12}{3} < \frac{3a}{3}$$

$$4 < a$$

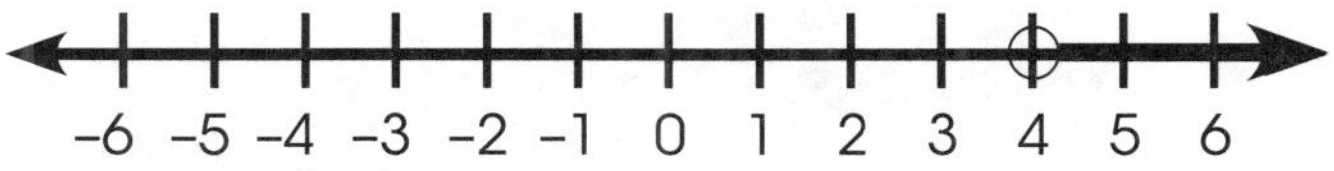

1. $4c + 1 < {}^{-}(5 + 2c)$

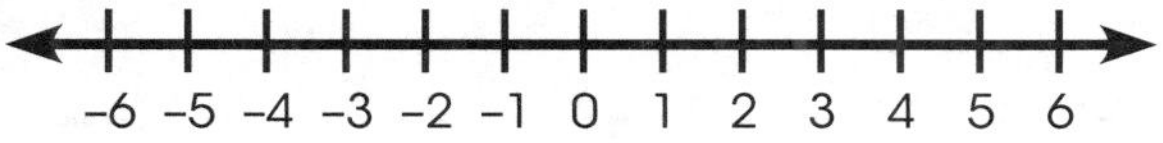

2. $2 - n > 2n + 11$

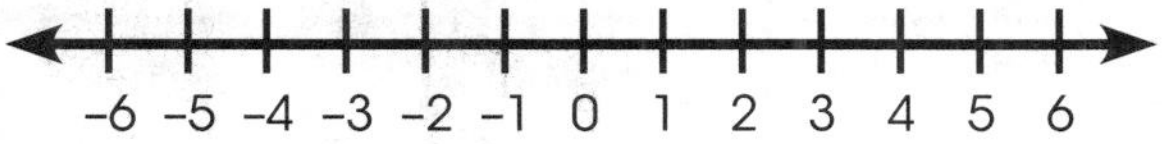

3. $2(3x - 5) > 2x + 6$

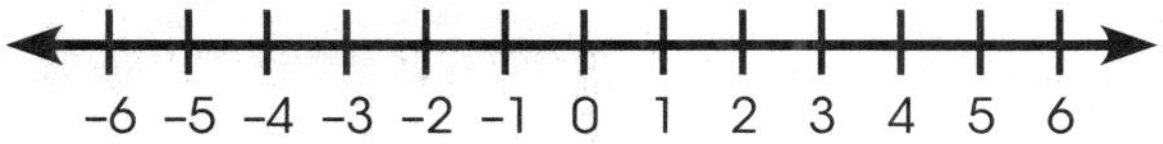

4. $^{-}2(4y - 21) \leq 12y - 16 + 9y$

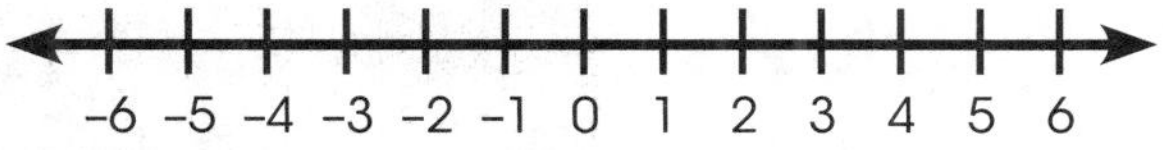

5. $n - 3n \geq {}^{-}4n - 7$

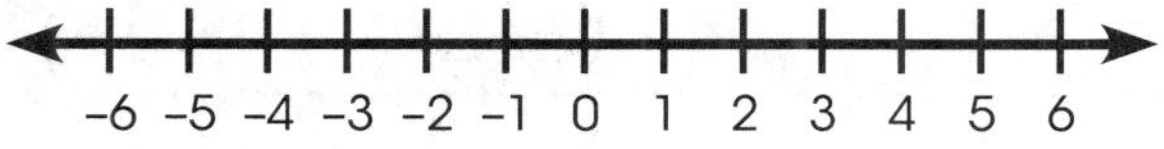

6. $10(x + 2) > {}^{-}2(6 - 9x)$

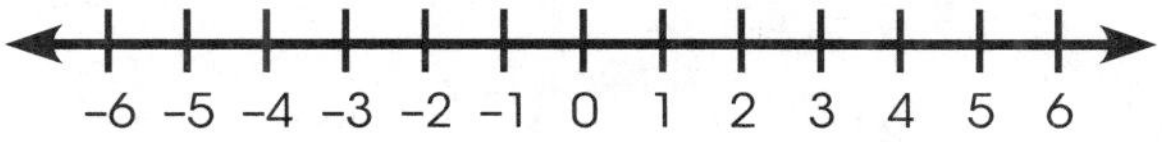

7. $11 + 3(^{-}8 + 5x) < 16x - 8$

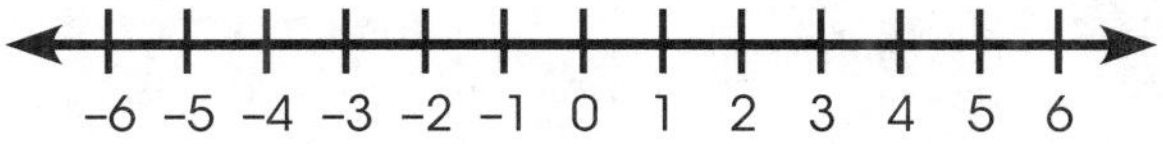

8. $12\ (2x + 3) \geq 3(9 + 7x)$

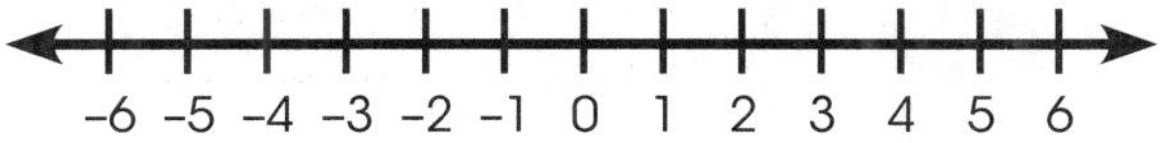

9. $35 - 18x > {}^{-}8(x + 3x)$

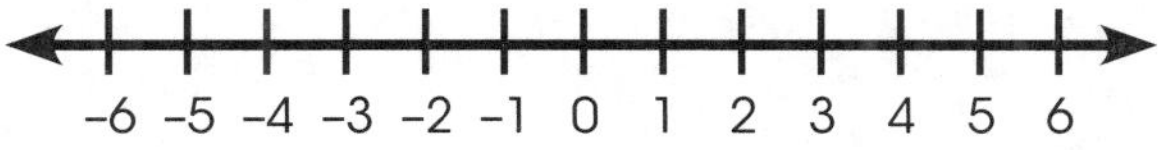

10. $12x + {}^{-}2(x + 5) < 3x(5 + 2) + 45$

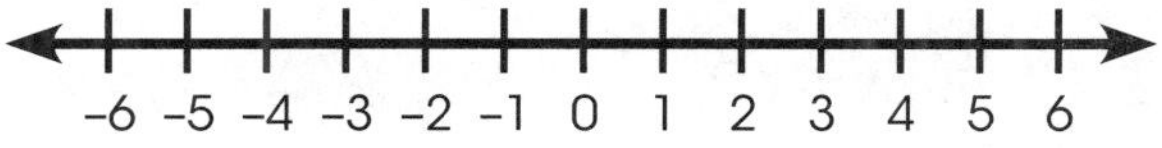

Name________________________________

7.EE.B.4b

Solving Multi-Step Inequalities

1. $32.4 \geq {}^{-}6c$

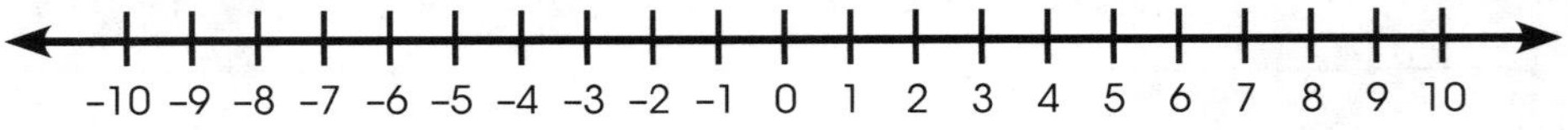

2. $x - {}^{-}15 \leq 9$

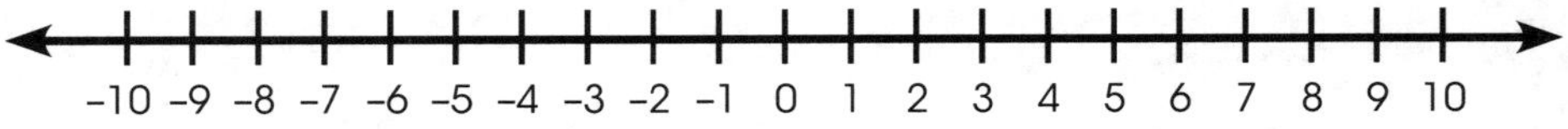

3. $-\frac{2}{3}b > {}^{-}6$

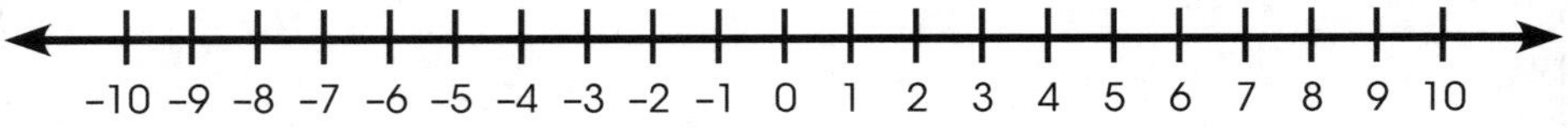

4. ${}^{-}18 + d > {}^{-}11$

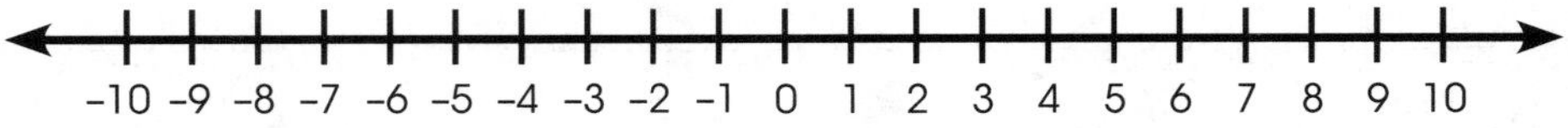

5. $6(2z + 3) \leq {}^{-}54$

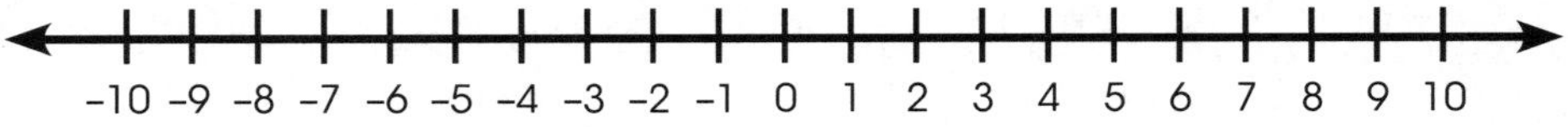

6. $8y - 15 < 27 + 2y$

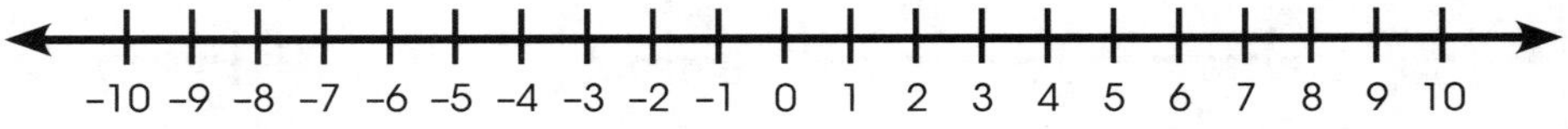

7. $162 > {}^{-}3a(5 + 1)$

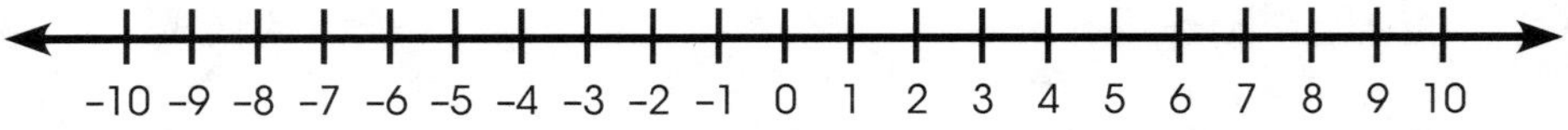

8. ${}^{-}6(5x + 8) \geq 2(8 - 7x)$

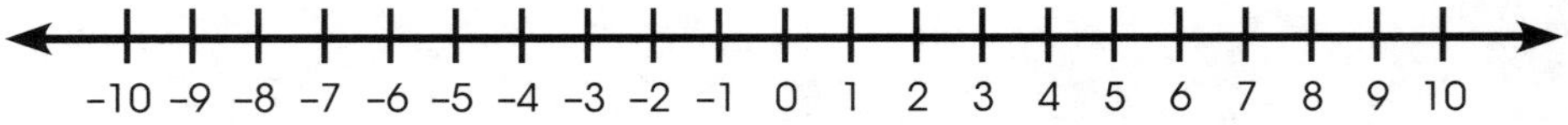

9. ${}^{-}40 \leq 8(2t - 2)$

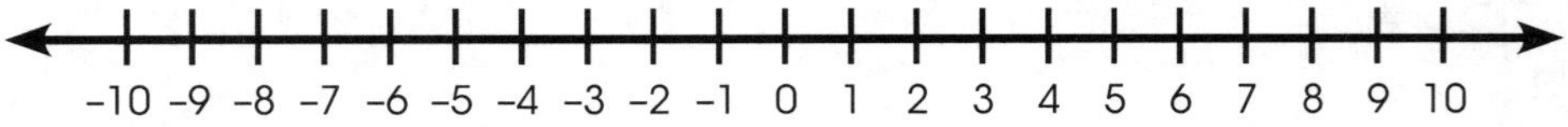

10. $5x(2 - 3) < 3x + 62$

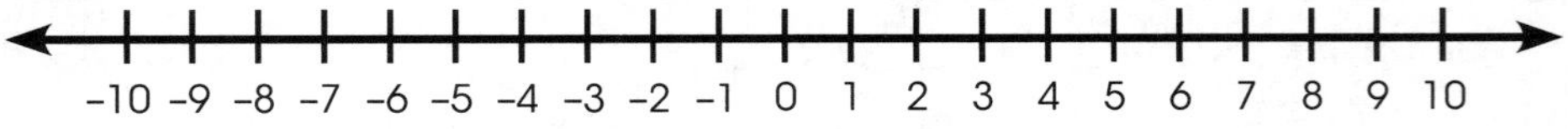

© Carson-Dellosa • CD-704384

Name________________________________

7.EE.B.4b

More Practice with Inequalities

1. $9x - 8 + x < 16 + 4x$

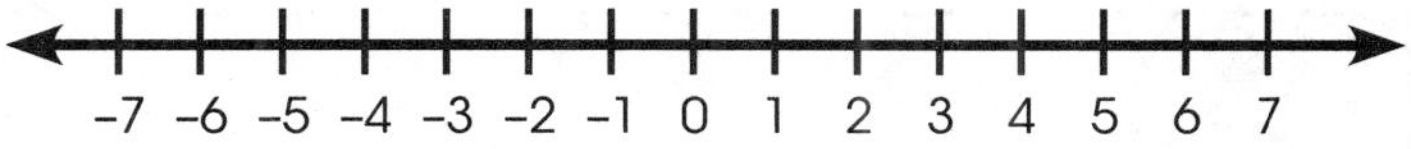

2. $15y \geq {}^{-}45$

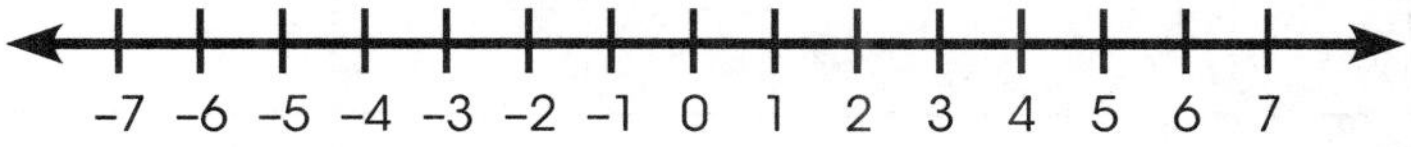

3. $69 > c + 71$

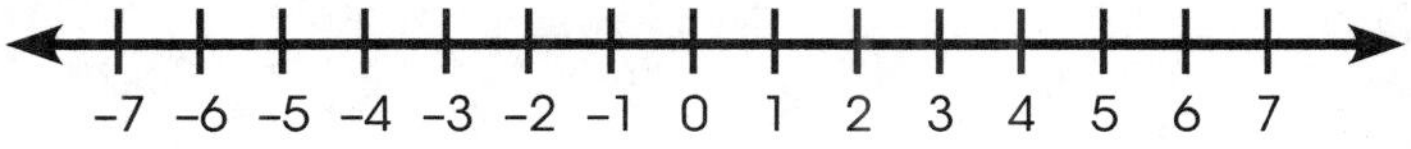

4. $17 + 11n - 13 \leq 4(n + 1) + 2n$

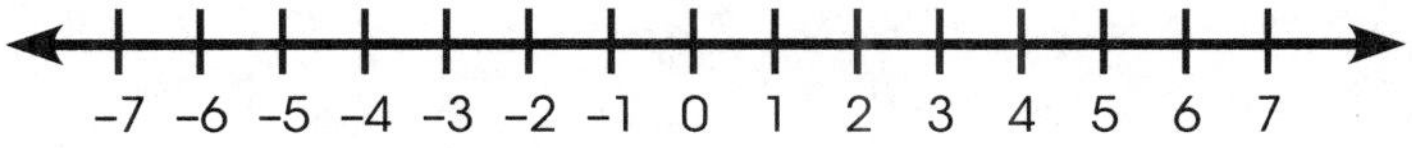

5. $8(2 + x) > 3(x - 3)$

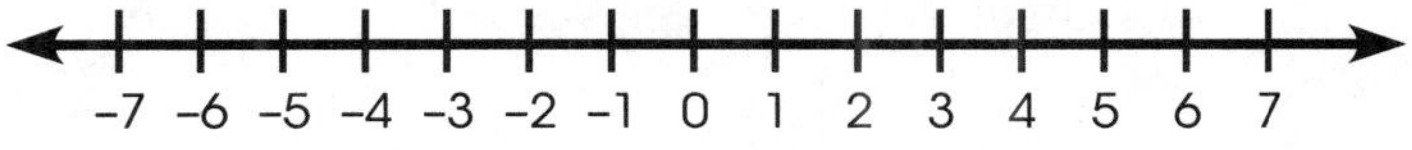

6. ${}^{-}4(3x + 2) \geq 40$

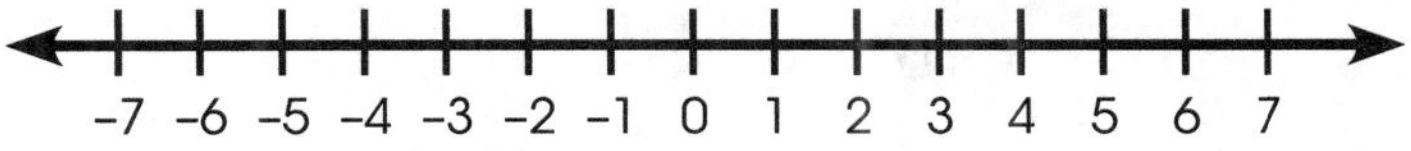

7. $\frac{5}{3} < \frac{2}{3}x - 1$

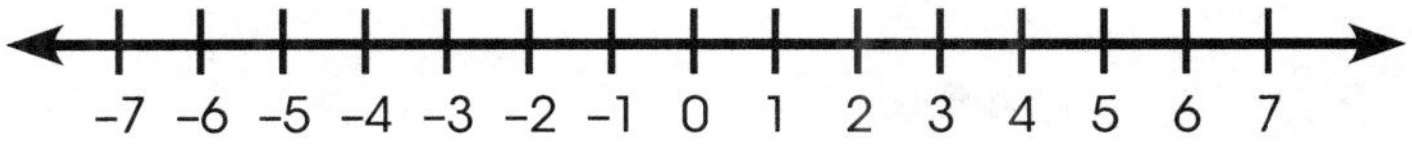

8. $3n - 4(2n - 5) + n + 4 \geq 0$

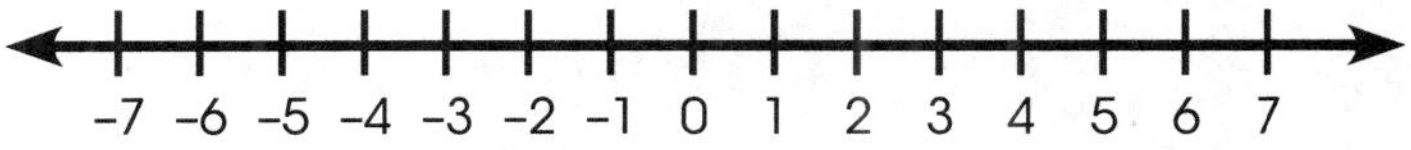

9. $18c + 11 - 26c < -3c(5 + 1) - 59$

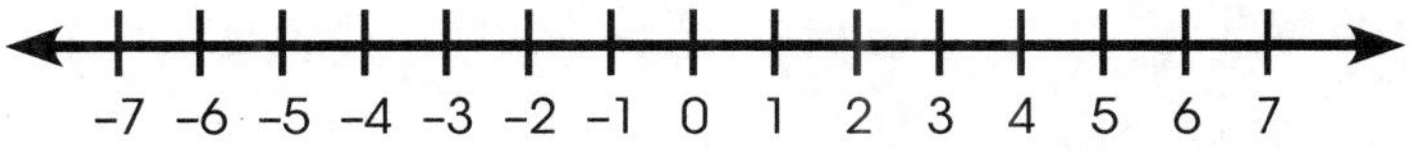

10. $8a - 2(2a + 5) \leq 2a(9 + 1) + 54$

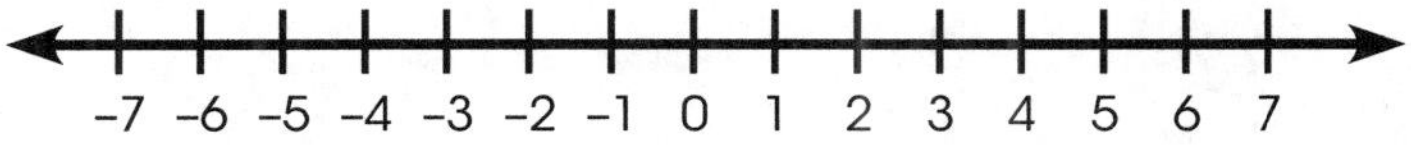

Name________________________________

7.EE.B.3

Solving Addition Equations

$$1.8 = {}^{-}2.1 + x$$
$$1.8 + 2.1 = {}^{-}2.1 + 2.1 + x$$
$$3.9 = 0 + x$$
$$3.9 = x$$

1. $a + {}^{-}7 = 8$

2. $y + 76 = {}^{-}93$

3. $4 + b = {}^{-}14$

4. ${}^{-}33 = z + 16$

5. ${}^{-}12 + x = 21$

6. $2.4 = m + 3.7$

7. ${}^{-}1\frac{1}{2} + n = {}^{-}1\frac{5}{8}$

8. ${}^{-}27 = c + 27$

9. $-\frac{5}{8} + x = -\frac{5}{8}$

10. $y + {}^{-}6.2 = 8.1$

11. $38 = x + {}^{-}19$

12. $a + {}^{-}2\frac{5}{9} = {}^{-}10\frac{5}{18}$

13. ${}^{-}1{,}129 + b = 3{,}331$

14. ${}^{-}3.5 = 7\frac{1}{2} + x$

© Carson-Dellosa • CD-704384

Name________________________________

7.EE.B.3

Solving Subtraction Equations

$$24 = x - {}^{-}8$$
$$24 = x + 8$$
$$24 - 8 = x + 8 - 8$$
$$16 = x + 0$$
$$16 = x$$

1. $k - 36 = 37$
2. $^{-}22 = y - 8$
3. $x - {}^{-}7 = {}^{-}19$
4. $30 = b - {}^{-}2$
5. $a - 18 = {}^{-}32$
6. $^{-}1.7 = b - 9.3$
7. $^{-}4\frac{1}{3} = q - 3\frac{1}{3}$
8. $^{-}17 = q - 3$
9. $p - \frac{3}{5} = \frac{3}{5}$
10. $5.62 = m - 6$
11. $x - {}^{-}36.5 = {}^{-}2.563$
12. $^{-}1{,}132 = b - 6{,}339$
13. $7\frac{3}{4} = a - 16\frac{3}{16}$
14. $z - {}^{-}5.75 = {}^{-}8\frac{1}{4}$

Name____ | 7.EE.B.3

Solving Addition and Subtraction Equations

1. $x + {}^{-}3 = -18$

2. $c - 11 = 43$

3. $12 + y = 32$

4. ${}^{-}26 = d - 7$

5. ${}^{-}62 = a + 16$

6. $q - {}^{-}83 = 121$

7. $t + {}^{-}101 = 263$

8. $w - 454 = {}^{-}832$

9. ${}^{-}332 = {}^{-}129 + s$

10. $665 = k - {}^{-}327$

11. ${}^{-}8.6 = m + 11.12$

12. $a - {}^{-}\frac{1}{5} = \frac{3}{20}$

13. $-\frac{3}{4} + z = \frac{7}{18}$

14. $b - 17.8 = {}^{-}36$

15. $-\frac{13}{24} = -\frac{5}{16} + c$

16. $102.8 = g - {}^{-}66.09$

17. $f + \frac{3}{5} = \frac{3}{4}$

18. $b - \frac{5}{6} = -\frac{7}{8}$

19. $21.21 + p = {}^{-}101.6$

20. ${}^{-}762.46 = h - 32.061$

© Carson-Dellosa • CD-704384

Name________________________________

7.EE.B.3

Solving Multiplication Equations

$$4y = {}^{-}28$$
$$\frac{4y}{4} = \frac{{}^{-}28}{4}$$
$$1y = {}^{-}7$$

1. ${}^{-}6a = {}^{-}66$

2. ${}^{-}180 = 12b$

3. ${}^{-}13n = 13$

4. $42 = {}^{-}14p$

5. $1\frac{1}{2} = 3x$

6. ${}^{-}5.6 = {}^{-}0.8x$

7. $8 = {}^{-}32b$

8. $9a = {}^{-}3$

9. $0.25y = 1.5$

10. ${}^{-}0.0006 = 0.02x$

11. ${}^{-}11x = 275$

12. $45\frac{1}{2} = {}^{-}14c$

13. $61.44 = 12z$

14. ${}^{-}21y = {}^{-}756$

Name________________________________

7.EE.B.3

Solving Division Equations

$$\frac{x}{4} = {}^-6$$

$$4 \cdot \frac{x}{4} = {}^-6 \cdot 4$$

$$x = {}^-24$$

1. ${}^-18 = \frac{a}{6}$

2. $\frac{x}{6} = {}^-6$

3. $\frac{y}{{}^-2} = 231$

4. $\frac{1}{5}b = {}^-8$

5. $\frac{m}{0.6} = 0.3$

6. $35 = \frac{x}{{}^-7}$

7. $0.12 = \frac{y}{0.12}$

8. $3 = -\frac{1}{8}a$

9. $\frac{w}{{}^-2} = 0.04$

10. $\frac{u}{{}^-4} = {}^-14$

11. $\frac{x}{{}^-5.1} = {}^-16$

12. $-28 = \frac{a}{13}$

13. $\frac{1}{18}c = {}^-31$

14. $\frac{b}{{}^-0.29} = 5.5$

© Carson-Dellosa • CD-704384

Name________________________________

7.EE.B.3

Solving Multiplication and Division Equations

1. ${}^{-}2p = {}^{-}38$

2. $\frac{b}{8} = {}^{-}24$

3. ${}^{-}85 = 17r$

4. ${}^{-}32 = \frac{c}{-22}$

5. ${}^{-}13a = 52$

6. $\frac{1}{47}d = {}^{-}26$

7. ${}^{-}12f = {}^{-}180$

8. $\frac{1}{0.16}x = 0.7$

9. ${}^{-}77.4 = 9a$

10. $-\frac{1}{6}q = {}^{-}11$

11. $16 = \frac{n}{{}^{-}21}$

12. $0.7h = {}^{-}0.112$

13. ${}^{-}80 = \frac{p}{15}$

14. $792 = {}^{-}33y$

15. ${}^{-}5.2 = \frac{m}{30.1}$

16. ${}^{-}11.2x = {}^{-}60.48$

17. $\frac{1}{{}^{-}26}r = {}^{-}66$

18. $315 = 21s$

19. $\frac{z}{0.06} = {}^{-}7.98$

20. ${}^{-}14g = {}^{-}406$

© Carson-Dellosa • CD-704384

Name______________________________ 7.EE.B.3

Solving Equations with Two Operations

$$2y - 7 = {}^{-}29$$
$$2y - 7 + 7 = {}^{-}29 + 7$$
$$2y = {}^{-}22$$
$$\frac{2y}{2} = \frac{{}^{-}22}{2}$$
$$y = {}^{-}11$$

1. $13 + {}^{-}3p = {}^{-}2$

2. $\frac{-5a}{2} = 75$

3. $6x - 4 = {}^{-}10$

4. $9 = 2y + 9$

5. ${}^{-}10 + \frac{a}{4} = 9$

6. $17 = 5 - x$

7. ${}^{-}7r - 8 = {}^{-}14$

8. $\frac{4y}{3} = 8$

9. $16 + \frac{x}{3} = {}^{-}10$

10. $\frac{-4z}{5} = -12$

11. ${}^{-}22 = 3s - {}^{-}8$

12. $-\frac{a}{6} - {}^{-}31 = 64$

© Carson-Dellosa • CD-704384

Name________________________________

7.EE.B.3

Magical Equations

Solve each equation. In a Magic Square, the sum of each row, column, and diagonal is the same.

1. $\frac{x}{4} + 5 = 7$	2. $2x - 20 = 10$	3. $-3x - 12 = 12$	4. $-x - 6 = -5$	5. $2 = 2x - 10$
6. $3x - 7 = 35$	7. $2 + 5x = -18$	8. $4x + 5 = -3$	9. $-11x + 10 = -45$	10. $5x - 6 = 29$
11. $-4 = \frac{4x}{5}$	12. $-2x + 7 = 13$	13. $\frac{5x}{4} + 2 = 7$	14. $-64 = -5x - 9$	15. $2x + 10 = 36$
16. $8x - 9 = -1$	17. $12 = 3x + 3$	18. $-4 = \frac{2x}{-5}$	19. $9 + 4x = 57$	20. $\frac{x}{-2} + 2 = 5$
21. $6 = \frac{x}{2} + 5$	22. $2x - 10 = 8$	23. $8 = \frac{x}{4} + 4$	24. $5x - 15 = -50$	25. $3x - 9 = -9$

The Magic Sum is ____________________.

Name______________________________ 7.EE.B.3

Mixed Practice

1. $4x - 7 = 2x + 15$

2. $^{-}4 = {}^{-}4(f - 7)$

3. $5x - 17 = 4x + 36$

4. $3(k + 5) = {}^{-}18$

5. $y + 3 = 7y - 21$

6. $^{-}3(m - 2) = 12$

7. $18 + 4p = 6p + 12$

8. $^{-}8\ (\frac{a}{8} - 2) = 26$

9. $^{-}3k + 10 = k + 2$

10. $22 = 2\ (b + 3)$

11. $6a + 9 = {}^{-}4a + 29$

12. $^{-}22 = 11(2c + 8)$

13. $10p - 14 = 9p + 17$

14. $^{-}45 = 5(\frac{2a}{5} + {}^{-}3)$

15. $16z - 15 = 13z$

16. $36 + 19b = 24b + 6$

17. $144 = {}^{-}16\ (3 + 3d)$

18. $11h - 14 = 7 + 14h$

19. $^{-}3\ (\frac{2j}{3} - 6) = 32$

20. $^{-}43 - 3z = 2 - 6z$

© Carson-Dellosa • CD-704384

Name________________________________

7.EE.B.3

Equation Steps

Solve these equations.

1. $^{-}116 = -a$

2. $6m - 2 = m + 13$

3. $x + 2 = {}^{-}61$

4. $^{-}18 = {}^{-}6 - y$

5. $^{-}5t + 16 = {}^{-}59$

6. $4a - 9 = 6a + 7$

7. $\frac{-3b}{8} = {}^{-}36$

8. $^{-}40 = 10\ (4 + s)$

9. $28 - \frac{k}{3} = 16$

10. $^{-}9r = 20 + r$

11. $114 = 11c - {}^{-}26$

12. $^{-}38 = 17 - 5z$

13. $^{-}5(2x - 5) = {}^{-}35$

14. $20c + 5 = 5c + 65$

15. $\frac{^{-}d}{5} - 21 = {}^{-}62$

16. $\frac{^{-}15c}{^{-}4} = {}^{-}30$

17. $384 = 12\ ({}^{-}8 + 5t)$

18. $3n + 7 = 7n - 13$

19. $^{-}8 - \frac{y}{3} = 22$

20. $^{-}5t - 30 = {}^{-}60$

HINT: The sum of the solutions equals the number of steps in the Statue of Liberty — 354!

Name________________________________ 6.EE.A.2a, 6.EE.B.6, 7.EE.B.3, 7.EE.B.4a

Solving Two-Step Problems

Write an equation and solve.

Ten more than 4 times a number is 6.
What is the number?

$$10 + 4n = 6$$

$$10 - 10 + 4n = 6 - 10$$

$$\frac{4n}{4} = \frac{^-4}{4}$$

$$n = {}^-1$$

1. Three-fifths of a number decreased by one is twenty-three. What is the number?

2. Seven more than six times a number is negative forty-seven. What is the number?

3. Nine less than twice a number is thirty-one. What is the number?

4. Three times the sum of a number and five times the number is thirty-six. What is the number?

5. The quotient of a number and four decreased by ten is two. What is the number?

6. Carol is sixty-six inches tall. This is twenty inches less than two times Mindy's height. How tall is Mindy?

7. In February, Paul's electric bill was three dollars more than one-half his gas bill. If the electric bill was ninety-two dollars, what was the gas bill?

© Carson-Dellosa • CD-704384

Name____________________________________

6.EE.C.9, 7.EE.B.3, 7.EE.B.4a

Solving Multi-Step Problems

Write an equation and solve.

One number is seven times a second number.

Their sum is 112. Find the numbers.

$$n + 7n = 112$$
$$\frac{8n}{8} = \frac{112}{8}$$
$$n = 14 \text{ and } 98$$

1. One of two numbers is five more than the other. The sum of the numbers is 17. Find the numbers.

2. The sum of two numbers is twenty-four. The larger number is three times the smaller number. Find the numbers.

3. One of two numbers is two-thirds the other number. The sum of the numbers is 45. Find the numbers.

4. The difference of two numbers is 19. The larger number is 3 more than twice the smaller. Find the numbers.

5. 320 tickets were sold to the school play. There were three times as many student tickets sold as adult tickets. Find the number of each.

6. The first number is eight more than the second number. Three times the second number plus twice the first number is equal to 26. Find the numbers.

7. Dan has five times as many \$1 bills as \$5 bills. He has a total of 48 bills. How many of each does he have?

Answer Key

1. 196 to 7	28	8. 54 out of 87	$\frac{18}{29}$
2. 19 : 76	$\frac{1}{4}$	9. 112 : 140	$\frac{4}{5}$
3. 18 out of 27	$\frac{2}{3}$	10. 88 to 104	$\frac{11}{13}$
4. $\frac{3}{8}$ to $\frac{3}{4}$	$\frac{1}{2}$	11. 65 out of 105	$\frac{13}{21}$
5. 0.11 : 1.21	$\frac{1}{11}$	12. 65 : 117	$\frac{5}{9}$
6. 140 : 112	$\frac{5}{4}$	13. 165 to 200	$\frac{33}{40}$
7. 18 to 27	$\frac{2}{3}$	14. 168 : 264	$\frac{7}{11}$

7

1. 55¢ to $4 $\frac{11}{80}$
2. 10 inches to 1 yard $\frac{5}{18}$
3. 8 hours to 3 days $\frac{1}{9}$
4. The ratio of wins to losses in 35 games with 21 losses and no ties $\frac{2}{3}$
5. The ratio of the area of a rectangle with sides of 6m and 8m to the area of a square with sides of length 12m $\frac{1}{3}$
6. The ratio of girls to boys in a class of 40 students with 17 girls $\frac{17}{23}$
7. Big Bob's batting average if he had 3 hits in 4 at bats against the Cougars $\frac{3}{4}$
8. The ratio of wins to losses in 42 games with 35 wins and no ties 5/[illegible]
9. A 36cm segment is divided into three parts whose lengths have the ratio of 2:3:7. Find the length of each segment. 6 cm, 9 cm, 21 cm
10. The sum of the measures of two complementary angles is 90°. Find the measures of two complementary angles whose measures are in the ratio of 1:4. 18°, 72°

8

1. $\frac{8}{6} = \frac{m}{27}$	36	9. $\frac{18}{15} = \frac{6}{x}$	5
2. $\frac{z}{3} = \frac{8}{15}$	$\frac{8}{5}$	10. $\frac{121}{x} = \frac{220}{100}$	55
3. $\frac{16}{40} = \frac{24}{c}$	60	11. $\frac{1.6}{x} = \frac{14}{21}$	2.4
4. $\frac{9}{p} = \frac{5}{2}$	$\frac{18}{5}$	12. $\frac{x}{168} = \frac{66\frac{2}{3}}{100}$	112
5. $\frac{1.8}{x} = \frac{3.6}{2.4}$	1.2	13. $\frac{x}{32} = \frac{37\frac{1}{2}}{100}$	12
6. $\frac{4}{5} = \frac{0.8}{y}$	1	14. $\frac{16}{48} = \frac{x}{100}$	$33\frac{1}{3}$
7. $\frac{x}{2} = \frac{15}{5}$	6	15. $\frac{0.12}{.25} = \frac{x}{100}$	48
8. $\frac{18}{12} = \frac{24}{x}$	16	16. $\frac{1.5}{x} = \frac{0.07}{0.14}$	3

9

1. If 64 feet of rope weigh 20 pounds, how much will 80 feet of the same type of rope weigh? **25 pounds**
2. If a 10 pound turkey takes 4 hours to cook, how long will it take a 14 pound turkey to cook? **5.6 hours**
3. An 18 ounce box of cereal costs $2.76. How many ounces should a box priced at $2.07 contain? **13.5 ounces**
4. Mike and Pat traveled 392 miles in 7 hours. If they travel at the same rate, how long will it take them to travel 728 miles? **13 hours**
5. If 2 pounds of turkey costs $1.98, what should 3 pounds cost? **$2.97**
6. If 2 liters of fruit juice cost $3.98, how much do 5 liters cost? **$9.95**
7. A 12 ounce box of cereal costs $.84. How many ounces should be in a box marked $.49? **7 ounces**
8. Janie saw an advertisement for a 6 ounce tube of toothpaste that costs $.90. How much should a 4 ounce tube cost? **$.60**

10

© Carson-Dellosa • CD-704384

Answer Key

11

1. $\frac{3}{4} = \frac{12}{A}$ 16
2. $\frac{4}{5} = \frac{C}{15}$ 12
3. $\frac{5}{D} = \frac{20}{10}$ 2.5
4. $\frac{E}{21} = \frac{2}{6}$ 7
5. $\frac{F}{4} = \frac{7}{8}$ 3.5
6. $\frac{3}{I} = \frac{30}{7}$ 0.7
7. $\frac{15}{L} = \frac{20}{L + 0.5}$ 1.5
8. $\frac{M}{17} = \frac{2}{4}$ 8.5
9. $\frac{4}{2.8} = \frac{3}{N}$ 2.1
10. $\frac{P}{3} = \frac{1}{15}$ 0.2
11. $\frac{Q}{18} = \frac{5}{4}$ 22.5
12. $\frac{7}{2} = \frac{R}{R - 15}$ 21
13. $\frac{5}{S} = \frac{4}{S - 2.5}$ 12.5
14. $\frac{18}{60} = \frac{T}{5T - 1.5}$ 0.9
15. If four cans of stew cost $5.00, what will be the unit (U) cost of 1 can? 1.25
16. Jamie rode her bike 52.5 miles east in 3 hours. What was her velocity (V) per hour? 15
17. Tony ran for a total of 75 yards during 6 games. At this rate, how many yards (Y) will Tony run in the remaining 2 games? 25

D○N'T PLACE Y○UR MEANS
2.5 2.1 0.9 0.2 1.5 16 12 7 25 1.25 21 8.5 7 16 2.1 12.5

○F C○NVEYANCE IN
3.5 12 2.1 17.5 7 25 16 2.1 12 7 0.7 2.1

FR○NT ○F Y○UR EQUINE
3.5 21 2.1 0.9 3.5 25 1.25 21 7 22.5 1.25 0.7 2.1 7

ANIMAL.
16 2.1 0.7 8.5 16 1.5

Write the familiar proverb.
Don't put the cart before the horse.

12

1.	$\frac{4}{5}$	80%	10.	11.3	1130%
2.	$\frac{4}{7}$	57.14%	11.	$\frac{11}{20}$	55%
3.	0.22	22%	12.	0.086	8.6%
4.	2.5	250%	13.	$\frac{7}{8}$	87.5%
5.	$\frac{3}{8}$	37.5%	14.	16.688	1668.8%
6.	0.006	0.6%	15.	$\frac{7}{16}$	43.75%
7.	1.125	112.5%	16.	621.9	62,190%
8.	$\frac{1}{2}$	50%	17.	$\frac{5}{16}$	31.25%
9.	$\frac{9}{40}$	22.5%	18.	3.9932	399.32%

13

80% of 30 =

$\frac{80}{100} = \frac{x}{30}$

$100x = 2400$

$x = 24$

1. 20% of 10 = 2
2. 25% of 45 = 11.25
3. 88% of 15 = 13.2
4. $9\frac{1}{2}$% of 20 = 1.9
5. 25% of 39 = 9.75
6. 16% of 90 = 14.4

___% of 30 = 10

$\frac{x}{100} = \frac{10}{40}$

$40x = 1000$

$x = 25$ 25%

1. 60 % of 25 = 15
2. $33\frac{1}{3}$ % of 30 = 10
3. 175 % of 4 = 7
4. 44 % of 75 = 33
5. 40 % of 15 = 6
6. 50 % of 80 = 40

80% of ___ = 65

$\frac{50}{100} = \frac{65}{x}$

$50x = 6500$

$x = 130$

1. 20% of 75 = 15
2. 80% of 70 = 56
3. 25% of 76 = 19
4. $33\frac{1}{3}$% of 123 = 41
5. 80% of 20 = 16
6. 30% of 50 = 15

14

1. In a group of 60 children, 12 have brown eyes. What percent have brown eyes? 20%
2. A salesman makes a 5% commission on all he sells. How much does he have to sell to make $1500? $30,000
3. A sales tax of $5\frac{3}{4}$ % is charged on a blouse priced at $42. How much sales tax must be paid? $2.42
4. A baby weighed 7.6 pounds at birth and $9\frac{1}{2}$ pounds after 6 weeks. What was the percent increase? 25%
5. A scale model of a building is 8% of actual size. If the model is 1.2 meters tall, how tall is the building? 15 meters
6. The purchase price of a camera is $84. The carrying case is 12% of the purchase price. Find the total cost including the carrying case. $94.08
7. The regular price of a record cost is $15. Find the discount and the new price if there is a 20% discount. discount = $3 new price = $12
8. A basketball team played 45 games. They won 60% of them. How many did the team win? 27
9. A test had 50 questions. Joe got 70% of them correct. How many did Joe get correct? 35
10. Diet soda contains 90% less calories than regular soda. If a can of regular soda contains 112 calories, how many calories does a can of diet soda contain? 11.2

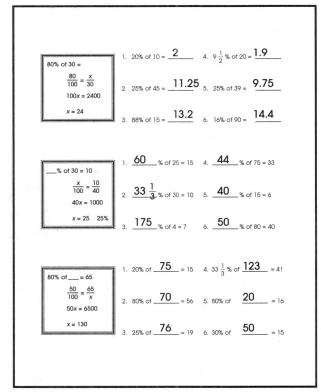

Answer Key

1. 72% of 310 — 223.2
2. 21 is 35% of what number? — 60
3. 28 out of 70 is what percent? — 40%
4. 6% of what number is 2.36? — $39\frac{1}{3}$
5. 3.9 is what percent of 10? — 39%
6. 115% of 12 — 13.8%
7. 60% of what number is 54? — 90
8. 17% of 800 is what number? — 136
9. What percent of 72 is 27? — 37.5
10. A piece of jewelry costs $78. If the price increases by 12%, what is the new cost? $87.36
11. Tax on a $24 item is $1.56. What is the tax rate (percent)? 6.5%
12. A dress was reduced in price by $19.56. This was 20% of the original price. Find the sale price. $78.24
13. There are 252 students on the student council at West High School. If there are 700 students enrolled, what percent are on the student council? 36%
14. One day 3% of the sweatshirts (or 15 sweatshirts) made at a factory were defective. How many sweatshirts were produced at the factory that day? 500

15

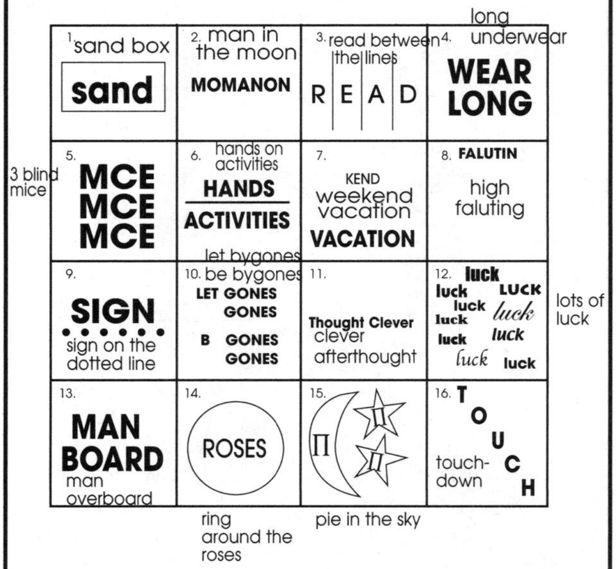

1. sand box
sand
2. man in the moon
MOMANON
3. read between the lines
R E A D
4. long underwear
WEAR LONG
5. 3 blind mice
MCE MCE MCE
6. hands on activities
HANDS
ACTIVITIES
7. weekend vacation
KEND
VACATION
8. FALUTIN
high faluting
9. SIGN
sign on the dotted line
10. let bygones be bygones
LET GONES
GONES
B GONES
GONES
11. Thought Clever
clever afterthought
12. luck LUCK luck luck luck luck luck luck luck
lots of luck
13. MAN BOARD
man overboard
14. ROSES
ring around the roses
15. pie in the sky
16. T O U C H
touch-down

16

1. $\frac{2}{9}+\frac{5}{9}=$ $\frac{7}{9}$
2. $\frac{3}{4}-\frac{1}{4}=$ $\frac{1}{2}$
3. $\frac{9}{15}+\frac{5}{15}=$ $\frac{14}{15}$
4. $\frac{19}{20}-\frac{14}{20}=$ $\frac{1}{4}$
5. $\frac{27}{38}+\frac{13}{38}=$ $1\frac{1}{19}$
6. $\frac{35}{60}-\frac{17}{60}=$ $\frac{3}{10}$
7. $\frac{17}{20}+\frac{23}{20}=$ 2
8. $\frac{25}{13}-\frac{12}{13}=$ 1
9. $\frac{11}{18}+\frac{16}{18}=$ $1\frac{1}{2}$
10. $\frac{17}{48}-\frac{14}{48}=$ $\frac{1}{16}$
11. $\frac{7}{45}+\frac{8}{45}=$ $\frac{1}{3}$
12. $\frac{33}{50}-\frac{17}{50}=$ $\frac{8}{25}$
13. $\frac{16}{33}+\frac{21}{33}=$ $1\frac{4}{33}$
14. $\frac{43}{56}-\frac{19}{56}=$ $\frac{3}{7}$
15. $\frac{12}{42}+\frac{31}{42}=$ $1\frac{1}{42}$
16. $\frac{29}{52}-\frac{13}{52}=$ $\frac{4}{13}$
17. $\frac{15}{18}+\frac{8}{18}=$ $1\frac{5}{18}$
18. $\frac{43}{65}-\frac{28}{65}=$ $\frac{3}{13}$

17

1. $\frac{2}{3}+\frac{5}{9}=$ $1\frac{2}{9}$
2. $\frac{4}{5}-\frac{3}{4}=$ $\frac{1}{20}$
3. $\frac{5}{6}+\frac{7}{12}=$ $1\frac{5}{12}$
4. $\frac{11}{15}-\frac{2}{5}=$ $\frac{1}{3}$
5. $\frac{11}{12}+\frac{5}{8}=$ $1\frac{13}{24}$
6. $\frac{1}{2}-\frac{4}{9}=$ $\frac{1}{18}$
7. $\frac{13}{36}+\frac{5}{12}=$ $\frac{7}{9}$
8. $\frac{7}{8}-\frac{3}{10}=$ $\frac{23}{40}$
9. $\frac{5}{12}-\frac{5}{18}=$ $\frac{5}{36}$
10. $\frac{5}{9}+\frac{3}{8}=$ $\frac{67}{72}$
11. $\frac{5}{12}-\frac{3}{15}=$ $\frac{13}{60}$
12. $\frac{3}{4}+\frac{7}{12}=$ $1\frac{1}{3}$
13. $\frac{8}{19}-\frac{1}{3}=$ $\frac{5}{57}$
14. $\frac{7}{15}+\frac{3}{25}=$ $\frac{44}{75}$
15. $\frac{30}{36}-\frac{5}{18}=$ $\frac{5}{9}$
16. $\frac{4}{5}+\frac{12}{13}=$ $1\frac{47}{65}$

18

© Carson-Dellosa • CD-704384

Answer Key

1.	$\frac{4}{9}+\frac{13}{15}=$	$1\frac{14}{45}$	11.	$\frac{11}{12}-\frac{1}{18}=$	$\frac{31}{36}$
2.	$\frac{5}{6}+\frac{7}{32}=$	$1\frac{5}{96}$	12.	$\frac{7}{23}-\frac{1}{7}=$	$\frac{26}{161}$
3.	$\frac{13}{15}-\frac{1}{3}=$	$\frac{8}{15}$	13.	$\frac{8}{21}+\frac{36}{49}=$	$1\frac{17}{147}$
4.	$\frac{3}{11}+\frac{6}{7}=$	$1\frac{10}{77}$	14.	$\frac{7}{9}-\frac{1}{4}=$	$\frac{19}{36}$
5.	$\frac{5}{9}-\frac{1}{15}=$	$\frac{22}{45}$	15.	$\frac{11}{30}+\frac{2}{25}=$	$\frac{67}{150}$
6.	$\frac{7}{9}+\frac{1}{6}=$	$\frac{17}{18}$	16.	$\frac{27}{35}-\frac{11}{30}=$	$\frac{17}{42}$
7.	$\frac{9}{10}-\frac{3}{20}=$	$\frac{3}{4}$	17.	$\frac{7}{8}+\frac{13}{14}=$	$1\frac{45}{56}$
8.	$\frac{11}{42}+\frac{1}{7}=$	$\frac{17}{42}$	18.	$\frac{76}{81}-\frac{22}{63}=$	$\frac{334}{567}$
9.	$\frac{8}{9}-\frac{1}{12}=$	$\frac{29}{36}$	19.	$\frac{1}{3}+\frac{2}{3}=$	1
10.	$\frac{7}{12}+\frac{31}{42}=$	$1\frac{9}{28}$	20.	$\frac{23}{45}-\frac{1}{3}=$	$\frac{8}{45}$

~~$\frac{19}{36}$ CAN~~	$\frac{17}{24}$ PUT	~~$\frac{17}{18}$ IT~~	~~$\frac{334}{567}$ PLACE~~	$\frac{7}{8}$ FORTH	~~1 IS~~	~~$1\frac{14}{45}$ WAS~~	~~$1\frac{17}{147}$ WHOLE~~	~~$\frac{29}{36}$ IT~~	$\frac{5}{27}$ HALF	~~$1\frac{5}{56}$ PROPER~~
$\frac{3}{8}$ THE	~~$\frac{26}{161}$ THIS~~	~~$\frac{17}{42}$ ON~~	~~$\frac{31}{36}$ TRY~~	$1\frac{35}{68}$ EFFORT	$\frac{31}{80}$ AND	~~$\frac{8}{15}$ SUM~~	$\frac{9}{30}$ YOU	$\frac{11}{45}$ GET	$\frac{1}{4}$ A	$1\frac{47}{147}$ FRACTION
~~$\frac{8}{45}$ IF~~	~~$\frac{17}{42}$ ARE~~	~~$\frac{22}{45}$ IN~~	$\frac{13}{55}$ OF	~~$1\frac{10}{77}$ TOTAL~~	~~$\frac{67}{150}$ HAS~~	$\frac{3}{5}$ THE	~~$1\frac{45}{56}$ VALUE~~	~~$1\frac{9}{28}$ TO~~	~~$\frac{3}{4}$ ALL~~	$1\frac{7}{47}$ RESULTS

PUT FORTH HALF THE EFFORT AND
YOU GET A FRACTION OF THE RESULTS

19

1.	$1\frac{1}{4}+2\frac{1}{2}=$	$3\frac{3}{4}$	9.	$1\frac{5}{6}+4=$	$5\frac{5}{6}$
2.	$5\frac{7}{10}-1\frac{1}{6}=$	$4\frac{8}{15}$	10.	$6\frac{7}{9}-6\frac{1}{2}=$	$\frac{5}{18}$
3.	$8\frac{3}{8}+9\frac{2}{3}=$	$18\frac{1}{24}$	11.	$7\frac{1}{4}+1\frac{7}{9}+2\frac{5}{6}=$	$11\frac{31}{36}$
4.	$6-2\frac{8}{11}=$	$3\frac{3}{11}$	12.	$8\frac{1}{6}-7\frac{3}{4}=$	$\frac{5}{12}$
5.	$2\frac{1}{16}+2\frac{1}{3}=$	$4\frac{19}{48}$	13.	$5+3\frac{3}{11}=$	$8\frac{3}{11}$
6.	$7\frac{7}{8}-7\frac{5}{12}=$	$\frac{11}{24}$	14.	$3\frac{5}{8}-1\frac{6}{7}=$	$1\frac{43}{56}$
7.	$4\frac{1}{2}+6\frac{2}{5}=$	$10\frac{9}{10}$	15.	$4\frac{3}{7}+5\frac{5}{14}=$	$9\frac{11}{14}$
8.	$5\frac{1}{2}-\frac{11}{15}=$	$4\frac{23}{30}$	16.	$6\frac{3}{12}-3\frac{9}{36}=$	3

20

I	1.	$7\frac{3}{5}+2\frac{1}{2}=$	$10\frac{1}{10}$	A.	$8\frac{1}{3}+12\frac{1}{2}=$	$20\frac{5}{6}$
D	2.	$10\frac{3}{5}-4=$	$6\frac{3}{5}$	B.	$2\frac{5}{18}-\frac{11}{24}=$	$1\frac{59}{72}$
E	3.	$5\frac{2}{9}+7\frac{1}{3}=$	$12\frac{5}{9}$	C.	$5\frac{2}{9}-3\frac{7}{9}=$	$1\frac{4}{9}$
G	4.	$11\frac{5}{6}-3\frac{3}{4}=$	$8\frac{1}{12}$	D.	$13-6\frac{2}{5}=$	$6\frac{3}{5}$
H	5.	$4\frac{7}{12}+4\frac{3}{14}=$	$8\frac{67}{84}$	E.	$3\frac{5}{6}+8\frac{13}{18}=$	$12\frac{5}{9}$
C	6.	$8-6\frac{5}{9}=$	$1\frac{4}{9}$	F.	$17\frac{5}{12}-4\frac{7}{12}=$	$12\frac{5}{6}$
A	7.	$17\frac{14}{15}+2\frac{9}{10}=$	$20\frac{5}{6}$	G.	$7\frac{11}{12}+\frac{1}{6}=$	$8\frac{1}{12}$
B	8.	$1\frac{17}{18}-\frac{1}{8}=$	$1\frac{59}{72}$	H.	$10\frac{13}{14}-2\frac{11}{84}=$	$8\frac{67}{84}$
F	9.	$6\frac{1}{12}+6\frac{3}{4}=$	$12\frac{5}{6}$	I.	$5\frac{1}{5}+4\frac{9}{10}=$	$10\frac{1}{10}$
J	10.	$8\frac{2}{9}-6\frac{17}{18}=$	$1\frac{5}{18}$	J.	$9\frac{1}{6}-7\frac{8}{9}=$	$1\frac{5}{18}$

21

1.	$\frac{1}{2}\cdot\frac{5}{6}=$	$\frac{5}{12}$	10.	$\frac{5}{6}\cdot 2=$	$1\frac{2}{3}$
2.	$3\cdot\frac{1}{2}=$	$1\frac{1}{2}$	11.	$8\frac{1}{3}\cdot\frac{3}{4}=$	$6\frac{1}{4}$
3.	$\frac{2}{5}\cdot\frac{1}{3}=$	$\frac{2}{15}$	12.	$4\frac{1}{4}\cdot 3\frac{1}{5}=$	$13\frac{3}{5}$
4.	$\frac{16}{5}\cdot\frac{25}{27}=$	$2\frac{26}{27}$	13.	$2\frac{1}{6}\cdot\frac{18}{20}=$	$1\frac{19}{20}$
5.	$\frac{8}{21}\cdot 2\frac{7}{16}=$	$\frac{13}{14}$	14.	$\frac{21}{35}\cdot 3\frac{4}{7}=$	$2\frac{1}{7}$
6.	$1\frac{5}{7}\cdot 2\frac{1}{4}=$	$3\frac{6}{7}$	15.	$1\frac{3}{5}\cdot 2\frac{3}{16}=$	$3\frac{1}{2}$
7.	$5\frac{7}{8}\cdot 4=$	$23\frac{1}{2}$	16.	$6\frac{3}{4}\cdot 1\frac{5}{9}=$	$10\frac{1}{2}$
8.	$\frac{5}{7}\cdot\frac{7}{5}=$	1	17.	$3\frac{1}{3}\cdot 1\frac{3}{18}=$	$3\frac{8}{9}$
9.	$3\frac{2}{3}\cdot\frac{17}{22}=$	$2\frac{5}{6}$	18.	$\frac{1}{2}\cdot\frac{6}{11}\cdot\frac{3}{5}=$	$\frac{9}{55}$

22

Answer Key

FRACTIONS QUIZ — Name: *Mort*

Problem	Answer
1. $5\frac{3}{5} \cdot 3\frac{4}{7} = 15\frac{12}{35}$	20
2. $2\frac{1}{12} \cdot 2\frac{2}{15} = 4\frac{4}{9}$	
3. $1\frac{1}{15} \cdot 3\frac{3}{7} = 3\frac{23}{35}$	
4. $8\frac{2}{9} \cdot 2\frac{7}{8} = 16\frac{7}{36}$	$23\frac{23}{36}$
5. $16 \cdot 4\frac{1}{4} = 68$	
6. $6\frac{2}{3} \cdot 1\frac{15}{16} = 12\frac{11}{12}$	
7. $5\frac{1}{3} \cdot 4\frac{1}{2} = 20\frac{1}{6}$	24
8. $9\frac{1}{3} \cdot 8\frac{1}{10} = 75\frac{3}{5}$	
9. $2\frac{1}{12} \cdot 3\frac{5}{9} = 7\frac{11}{27}$	
10. $3\frac{5}{6} \cdot 8 = 24\frac{5}{6}$	$30\frac{2}{3}$
11. $9\frac{1}{3} \cdot 1\frac{5}{7} \cdot \frac{3}{4} = 9\frac{15}{84}$	12
12. $6\frac{8}{9} \cdot 3\frac{6}{7} = 26\frac{4}{7}$	
13. $8\frac{2}{5} \cdot 3\frac{1}{3} = 24\frac{2}{15}$	28
14. $9\frac{3}{5} \cdot 2\frac{1}{12} = 20$	
15. $2\frac{1}{2} \cdot 2\frac{8}{9} = 4\frac{4}{9}$	$7\frac{2}{9}$
16. $5\frac{3}{7} \cdot 2\frac{3}{16} = 10\frac{9}{112}$	$11\frac{7}{8}$
17. $2\frac{1}{4} \cdot 6 \cdot 1\frac{1}{9} = 15$	
18. $4\frac{1}{2} \cdot 2\frac{2}{5} = 8\frac{1}{5}$	$10\frac{4}{5}$
19. $7\frac{1}{2} \cdot 7\frac{1}{3} = 55$	
20. $3\frac{1}{8} \cdot \frac{1}{9} \cdot \frac{9}{10} = 3\frac{1}{80}$	$\frac{5}{16}$

Write mixed numbers as fractions, x numerators, x denominators.

23

1. $\frac{3}{7} \div \frac{1}{2} =$	$\frac{6}{7}$	10. $\frac{7}{8} \div 2\frac{1}{3} =$	$\frac{3}{8}$
2. $\frac{17}{9} \div \frac{8}{9} =$	$2\frac{1}{8}$	11. $9\frac{3}{8} \div 3\frac{3}{4} =$	$2\frac{1}{2}$
3. $6\frac{2}{3} \div 5 =$	$1\frac{1}{3}$	12. $5\frac{1}{6} \div \frac{31}{6} =$	1
4. $1\frac{7}{9} \div 4\frac{2}{9} =$	$\frac{8}{19}$	13. $\frac{7}{8} \div \frac{3}{4} =$	$1\frac{1}{6}$
5. $\frac{15}{4} \div \frac{5}{14} =$	$10\frac{1}{2}$	14. $\frac{7}{12} \div \frac{7}{4} =$	$\frac{1}{3}$
6. $\frac{11}{12} \div \frac{13}{8} =$	$\frac{22}{39}$	15. $4\frac{6}{7} \div \frac{1}{3} =$	$14\frac{4}{7}$
7. $4 \div 4\frac{2}{5} =$	$\frac{10}{11}$	16. $5\frac{1}{2} \div \frac{7}{4} =$	$3\frac{1}{7}$
8. $3\frac{1}{4} \div 4\frac{3}{8} =$	$\frac{26}{35}$	17. $2\frac{2}{9} \div 4\frac{2}{6} =$	$\frac{20}{39}$
9. $\frac{6}{15} \div \frac{9}{10} =$	$\frac{4}{9}$	18. $5\frac{5}{12} \div 3\frac{1}{3} =$	$1\frac{5}{8}$

24

$\frac{5}{12} \div \frac{1}{2}$ $\frac{5}{6}$	$1\frac{1}{2} \div 1\frac{1}{3}$ $1\frac{1}{8}$	$\frac{5}{6} \div 5$ $\frac{1}{6}$	$\frac{11}{12} \div 2$ $\frac{11}{24}$	$1\frac{1}{2} \div 2$ $\frac{3}{4}$
$6\frac{1}{2} \div 6$ $1\frac{1}{12}$	$\frac{3}{4} \div \frac{9}{4}$ $\frac{1}{3}$	$\frac{5}{6} \div 2$ $\frac{5}{12}$	$2\frac{1}{8} \div 3$ $\frac{17}{24}$	$1\frac{7}{12} \div 2$ $\frac{19}{24}$
$\frac{7}{8} \div 3$ $\frac{7}{24}$	$\frac{6}{7} \div 2\frac{2}{7}$ $\frac{3}{8}$	$\frac{2}{5} \div \frac{3}{5}$ $\frac{2}{3}$	$2\frac{7}{8} \div 3$ $\frac{23}{24}$	$\frac{5}{12} \div \frac{2}{5}$ $1\frac{1}{24}$
$\frac{13}{48} \div \frac{1}{2}$ $\frac{13}{24}$	$\frac{5}{32} \div \frac{1}{4}$ $\frac{5}{8}$	$\frac{2}{3} \div \frac{8}{11}$ $\frac{11}{12}$	$\frac{7}{8} \div \frac{7}{8}$ 1	$\frac{1}{2} \div 2$ $\frac{1}{4}$
$1\frac{1}{2} \div 2\frac{4}{7}$ $\frac{7}{12}$	$1\frac{3}{4} \div 2$ $\frac{7}{8}$	$\frac{7}{18} \div \frac{1}{3}$ $1\frac{1}{6}$	$\frac{1}{3} \div 1\frac{3}{5}$ $\frac{5}{24}$	$\frac{3}{4} \div 1\frac{1}{2}$ $\frac{1}{2}$

Magic Sum = $3\frac{1}{3}$

If every row and column in a Magic Square of problems has the same sum except for the last row and the last column, what do you know?

The problem in the bottom right corner is wrong.

25

FRACTIONS QUIZ — Name: *Cal*

Problem	Answer
1. $\frac{3}{5} + \frac{1}{3} = \frac{2}{5}$	$\frac{14}{15}$
2. $\frac{3}{4} + \frac{3}{4} = \frac{6}{8}$	$1\frac{1}{2}$
3. $4\frac{2}{3} + 6\frac{3}{4} = 10\frac{5}{7}$	$11\frac{5}{12}$
4. $2\frac{1}{2} + 3\frac{1}{2} = 6$	
5. $\frac{7}{8} - \frac{2}{3} = \frac{5}{24}$	
6. $\frac{6}{7} - \frac{2}{7} = \frac{4}{7}$	
7. $2\frac{4}{5} - 1\frac{2}{3} = 1\frac{2}{15}$	
8. $6\frac{1}{4} - 2\frac{3}{4} = 4\frac{1}{2}$	$3\frac{1}{2}$
9. $\frac{3}{4} \cdot \frac{6}{7} = \frac{1}{14}$	$\frac{9}{14}$
10. $\frac{1}{3} \cdot \frac{1}{3} = \frac{1}{6}$	$\frac{1}{9}$
11. $1\frac{2}{3} \cdot 2\frac{1}{2} = 2\frac{1}{3}$	$4\frac{1}{6}$
12. $4\frac{1}{2} \cdot 3\frac{1}{3} = 12\frac{1}{6}$	15
13. $\frac{3}{4} \div \frac{1}{2} = 1\frac{1}{2}$	
14. $\frac{2}{3} \div \frac{3}{4} = \frac{8}{9}$	
15. $2\frac{4}{5} \div 1\frac{2}{5} = 2\frac{2}{5}$	2
16. $5\frac{1}{4} \div 3\frac{1}{2} = 15\frac{1}{8}$	$1\frac{1}{2}$

What rules for computing with fractions would you share with Cal?

Addition: Use a common denominator, add numerators.

Subtraction: Use a common denominator, subtract numerators.

Multiplication: Multiply numerators, multiply denominators.

Division: Invert and multiply.

* Write mixed numbers as simple fractions.

26

© Carson-Dellosa • CD-704384

Answer Key

1. $8\frac{1}{15} - 5\frac{11}{20} =$ $2\frac{31}{60}$
2. $3\frac{1}{9} + 8\frac{3}{7} + 1\frac{1}{3} =$ $12\frac{55}{63}$
3. $1\frac{7}{8} \bullet 3\frac{3}{5} =$ $6\frac{3}{4}$
4. $4\frac{4}{5} \div 2\frac{8}{10} =$ $1\frac{5}{7}$
5. $3\frac{5}{12} + 5\frac{1}{4} - 2\frac{7}{20} =$ $6\frac{19}{60}$
6. $(\frac{16}{21} \bullet 3\frac{1}{4}) + 6\frac{1}{3} =$ $8\frac{17}{21}$
7. $5\frac{7}{10} - (\frac{25}{27} \div 3\frac{1}{3}) =$ $5\frac{19}{45}$
8. $(2\frac{15}{24} + 3\frac{11}{12}) \bullet 6\frac{1}{2} =$ $42\frac{25}{48}$
9. $7\frac{3}{12} - 2\frac{8}{9} =$ $4\frac{13}{36}$
10. $1\frac{1}{6} \bullet 3\frac{5}{7} \bullet 2\frac{2}{9} =$ $9\frac{17}{27}$
11. $8\frac{7}{12} + 11\frac{3}{4} =$ $20\frac{1}{3}$
12. $7 - (3\frac{7}{9} \div 4\frac{2}{3}) =$ $6\frac{4}{21}$
13. $2\frac{1}{2} \bullet 3\frac{3}{15} =$ 8
14. $5\frac{2}{9} - 2\frac{17}{18} + 1\frac{2}{3} =$ $3\frac{17}{18}$
15. $(3\frac{6}{8} \div 4\frac{2}{4}) - \frac{13}{16} =$ $\frac{1}{48}$
16. $4\frac{2}{3} \bullet 1\frac{3}{4} \bullet 3\frac{3}{4} =$ $30\frac{5}{8}$
17. $3\frac{4}{15} + 8\frac{3}{45} =$ $11\frac{1}{3}$
18. $12\frac{1}{2} - 7\frac{15}{16} =$ $4\frac{9}{16}$
19. $(1\frac{12}{13} \bullet 7\frac{3}{5}) - 3 =$ $11\frac{8}{13}$
20. $2\frac{1}{8} + (6\frac{2}{3} \div 8\frac{4}{9}) =$ $2\frac{139}{152}$
21. $3\frac{1}{3} \bullet 7\frac{5}{6} \bullet 2\frac{2}{5} =$ $62\frac{2}{3}$
22. $1\frac{15}{16} + 3\frac{7}{24} + 3\frac{11}{12} =$ $9\frac{7}{48}$

27

1. If $1\frac{1}{4}$ pounds of bananas sell for 80¢ and $1\frac{1}{3}$ pounds of apples sell for 90¢, which fruit is cheaper? bananas 64¢/lb.
2. A cake recipe calls for $\frac{2}{3}$ teaspoons of salt, $1\frac{1}{2}$ teaspoons baking powder, 1 teaspoon baking soda and $\frac{1}{2}$ teaspoon cinnamon. How many total teaspoons of dry ingredients are used? $3\frac{2}{3}$ teaspoons
3. A baseball team played 35 games and won $\frac{4}{7}$ of them.
   How many games were won? 20
   How many games were lost? 15
4. During 4 days, the price of the stock of PEV Corporation went up $\frac{1}{4}$ of a point, down $\frac{1}{3}$ of a point, down $\frac{3}{4}$ of a point and up $\frac{7}{10}$ of a point. What was the net change? down $\frac{2}{15}$
5. Janie wants to make raisin cookies. She needs $8\frac{1}{2}$ cups of raisins for the cookies. A 15-ounce box of raisins contains $2\frac{3}{4}$ cups. How many boxes must Janie buy to make her cookies? 4 Boxes
6. A one-half gallon carton of milk costs $1.89. A one-gallon carton of milk costs $2.99. How much money would you save if you bought a one-gallon carton instead of 2 one-half gallon cartons? 79¢

28

1. $\frac{3}{5} =$ 0.6
2. $\frac{11}{25} =$ 0.44
3. $\frac{7}{15} =$ $0.4\overline{6}$
4. $2\frac{1}{9} =$ $2.\overline{1}$
5. $\frac{23}{33} =$ $\overline{.69}$
6. $1\frac{5}{16} =$ 1.3125
7. $\frac{12}{25} =$ 0.48
8. $\frac{1}{3} =$ $0.\overline{3}$
9. $\frac{5}{33} =$ $\overline{0.15}$
10. $2\frac{5}{16} =$ 2.3125
11. $\frac{25}{37} =$ $\overline{0.675}$
12. $3\frac{13}{15} =$ $3.8\overline{6}$
13. $\frac{17}{22} =$ $0.7\overline{72}$
14. $3\frac{11}{12} =$ $3.91\overline{6}$

29

Fraction	Decimal	T or R	Fraction	Decimal	T or R
1. $\frac{3}{8}$	0.375	T	11. $2\frac{3}{8}$	2.375	T
2. $\frac{8}{15}$	$0.5\overline{3}$	R	12. $2\frac{15}{37}$	$2.\overline{405}$	R
3. $\frac{27}{32}$	0.84375	T	13. $\frac{67}{90}$	$0.7\overline{4}$	R
4. $\frac{23}{30}$	$0.7\overline{6}$	R	14. $1\frac{19}{33}$	$1.\overline{57}$	R
5. $\frac{4}{7}$	$0.\overline{571428}$	R	15. $\frac{124}{333}$	$0.\overline{372}$	R
6. $5\frac{1}{8}$	5.125	T	16. $5\frac{7}{10}$	5.7	T
7. $1\frac{4}{5}$	1.8	T	17. $2\frac{11}{16}$	2.6875	T
8. $\frac{10}{35}$	$0.\overline{285714}$	R	18. $7\frac{31}{40}$	7.775	T
9. $\frac{9}{15}$	0.6	T	19. $3\frac{9}{16}$	3.5625	T
10. $2\frac{7}{8}$	2.875	T	20. $11\frac{3}{4}$	11.75	T

BONUS: For fractions in lowest terms, what are the prime factors of the denominators that terminate? 2 and 5

Give a rule for determining whether a fraction will be a terminating or repeating decimal.

Reduce to lowest terms. If the only factors of the denominator are 2 and/or 5, it terminates.

30

Answer Key

Round to the nearest whole number.

1. 41.803 = 42	2. 119.63 = 120	3. 20.05 = 20	4. 3.45 = 3
5. 79.531 = 80	6. 8.437 = 8	7. 29.37 = 29	8. 109.96 = 110

Round to the nearest tenth.

9. 33.335 = 33.3	10. 1.861 = 1.9	11. 99.96 = 100.0	12. 103.103 = 103.1
13. 16.031 = 16.0	14. 281.05 = 281.1	15. 8.741 = 8.7	16. 27.773 = 27.8

Round to the nearest hundredth.

17. 69.713 = 69.71	18. 5.569 = 5.57	19. 609.906 = 609.91	20. 247.898 = 247.90
21. 5.535 = 5.54	22. 67.1951 = 67.20	23. 14.0305 = 14.03	24. 6.9372 = 6.94

31

1.	4.81 x 100 = 481	10.	90,000 ÷ 100 = 900
2.	37.68 ÷ 10 = 3.768	11.	0.000618 x 1,000 = 0.618
3.	0.46 x 1,000 = 460	12.	39.006 ÷ 1,000 = 0.039006
4.	7.12 ÷ 10,000 = 0.000712	13.	16 x 100 = 1600
5.	5.4 x 10 = 54	14.	28.889 ÷ 10,000 = 0.0028889
6.	27,500 ÷ 1,000 = 27.5	15.	36.89 x 10,000 = 368900
7.	4.395 x 100,000 = 439,500	16.	0.091 ÷ 100 = 0.00091
8.	0.0075 ÷ 100 = 0.000075	17.	0.0336 x 100,000 = 3360
9.	2.274 x 10 = 22.74	18.	1,672 ÷ 100,000 = 0.01672

32

1. 3.5 + 8.4 =	11.9	11. 17.6 – 9.3 =	8.3
2. 43.57 + 104.6 =	148.17	12. 32.3 – 12.72 =	19.58
3. 15.36 + 29.23 + 7.2 =	51.79	13. 23.96 – 19.931 =	4.029
4. 2.304 + 6.18 + 9.2 =	17.684	14. $29.98 – $16.09 =	$13.89
5. $12.91 + $6.99 =	$19.90	15. 63.36 – 0.007 =	63.353
6. 0.08 + 19 =	19.08	16. 16.22 – 0.039 =	16.181
7. 22.63 + 1.694 =	24.324	17. 44.44 – 16.1 =	28.34
8. 362.1 + 8.888 + 0.016 =	371.004	18. $75.02 – $3.99 =	$71.03
9. 1392.16 + 16.16 =	1408.32	19. 575.021 – 65.98 =	509.041
10. 83.196 + 0.0017 =	83.1977	20. 394.6 – 27.88 – 0.0933 =	366.6267

33

1. 6.2 + 0.25 = 6.45	8. 77.7 – 7 = 70.7
2. 3.3 – 0.33 = 2.97	9. 7.8 + 64.2 = 72
3. 0.26 + 0.4 = 0.66	10. 9.25 – 2.5 = 6.75
4. 8.76 – 5.43 = 3.33	11. 36 + 6.3 = 42.3
5. 19.9 + 1.1 = 21	12. 37.2 – 32 = 5.2
6. 9.53 – 5.3 = 4.23	13. 0.23 + 3.7 = 3.93
7. 0.22 + 2.2 = 2.42	14. 28.55 – 20.5 = 8.05

15. 27.8 + 2.2 – 3.5 + 0.5 – 20.5 = 6.5

20 RE	~~3.33 AL~~	34 ME	~~70.7 WA~~	71 MB	~~6.75 YS~~	0.3 ER	~~6.45 CO~~	~~0.66 UN~~	9 TO	~~42.3 AD~~	~~3.93 DU~~	8.7 LI	~~8.05 PT~~
9.9 NE	~~2.42 HE~~	4.4 UP	~~5.2 PL~~	4.49 TH	77 EP	~~72 AC~~	0.6 OI	~~21 ES~~	~~4.23 TO~~	9 NT	~~2.97 AD~~	~~6.5 D.~~	3.07 S.

Write the remaining letters, one letter to space.

R E M E M B E R T O
L I N E U P T H E
P O I N T S

34

© Carson-Dellosa • CD-704384

Answer Key

35

1. (0.003) (6) = 0.018
2. (0.051) (0.003) = 0.000153
3. (260) (0.01) = 2.6
4. (9.6) (5) = 48.0
5. (7) (3.42) = 23.94
6. (5.29) (11.3) = 59.777
7. (0.017) (6.2) = 0.1054
8. (0.3) (0.03) (0.003) = 0.000027
9. (1.5) (0.096) (4.3) = .6192
10. (0.05) (0.16) (0.001) = 0.000008
11. (8) (0.217) (0.01) = 0.01736
12. (18) (0.08) = 1.44
13. (16.01) (0.5) (0.31) = 2.48155
14. (1.06) (0.005) = 0.0053
15. (4.802) (11.11) = 53.35022
16. (10.25) (0.331) = 3.39275
17. (5) (1.102) = 5.51
18. (12.8) (0.05) (3.09) = 1.9776

36

1. 2.2 x 0.011 = 242 — 0.0242
2. 12.8 x 0.12 = 1536 — 1.536
3. 1.8 x 6.03 = 10854 — 10.854
4. 34.1 x 1.4 = 4774 — 47.74
5. 7.21 x 22.2 = 160062 — 160.062
6. 55 x 0.033 = 1815 — 1.815
7. 6.9 x 11 = 759 — 75.9
8. 6.7 x 0.801 = 53667 — 5.3667
9. 4.04 x 4.04 = 163216 — 16.3216
10. 32.1 x 2.02 = 64842 — 64.842
11. 0.005 x 0.011 = 55 — 0.000055
12. 66.2 x 1.1 = 7282 — 72.82
13. 0.84 x 0.07 = 588 — 0.0588
14. 8.2 x 0.1 = 82 — 0.82
15. 0.6 x 1.7 = 102 — 1.02
16. (5.7) (0.2) (0.07) = 798 — 0.0798
17. (9.8) (2.8) (1.8) = 49392 — 49.392
18. (10.6) (4.3) (0.8) = 36464 — 36.464
19. (0.13) (8.5) (0.5) = 5525 — 0.5525
20. (6.7) (1.2) (0.03) = 2412 — 0.2412

HINT:
The sum of the number of all decimal places in your products equals 64.

37

1. 0.128 ÷ 0.8 = 0.16
2. 2.45 ÷ 3.5 = 0.7
3. 0.5773 ÷ 5.02 = 0.115
4. 39.78 ÷ 0.195 = 204
5. 4.2016 ÷ 5.2 = 0.808
6. 1.45 ÷ 0.08 = 18.125
7. 0.1716 ÷ 5.2 = 0.033
8. 3.906 ÷ 1.2 = 3.255
9. 6.56 ÷ 0.16 = 41
10. 0.0135 ÷ 4.5 = 0.003
11. 0.0483 ÷ 0.21 = 0.23
12. 0.5418 ÷ 0.3 = 1.806
13. 16.83 ÷ 0.11 = 153
14. 0.1926 ÷ 32.1 = 0.006

38

1. 12.16 – 8.72 = 3.44
2. 119.7 + 11.97 = 131.67
3. (3.4) (8) = 27.2
4. 2960 ÷ 0.37 = 8,000
5. 1.21 ÷ 1.1 = 1.1
6. 7 + 6.91 = 13.91
7. 18.91 – 11.857 = 7.053
8. (1.35) (21.4) = 28.89
9. 21.2 – 9.03 = 12.17
10. 0.7 + 0.02 + 4 = 4.72
11. (0.25) (2.5) (25) = 15.625
12. 95.6 – 87.81 +12.21 = 20
13. (0.8) (1.3) (0.62) = 0.6448
14. 37.92 ÷ 1.2 = 31.6
15. 0.1007 ÷ 5.3 = 0.019
16. 329.82 + 6.129 = 335.949
17. 893.631 – 11.09 = 882.541
18. 18.332 + 82.82 = 101.152
19. 132.03 ÷ 8.1 = 16.3
20. (16.1) (3.66) = 58.926
21. 1093.62 – 10.993 = 1082.627
22. 6.963 ÷ 2.11 = 3.3

Answer Key

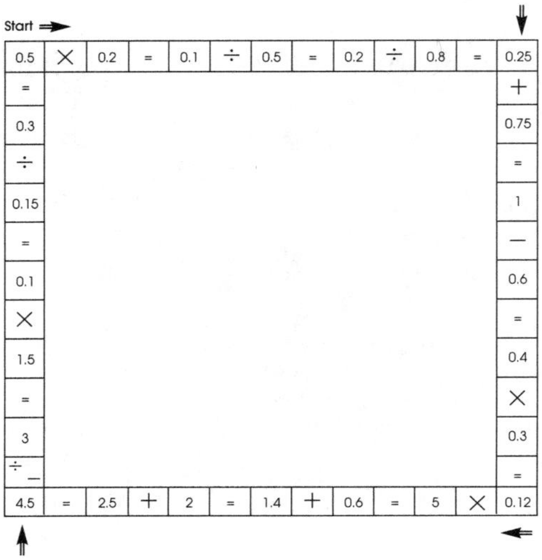

39

1. Jim's gas credit card bill was $80.97 for June, $41.35 for July and $65.08 for August. What were his total charges for the summer?
   $187.40
2. One cup of hot chocolate can be made with .18 ounces of hot chocolate mix. How many cups can be made from a 6.48 ounce canister of mix?
   36 cups
3. Karl's car payments are $215.37 per month for the next three years. What will be the total amount he will pay for his car?
   $7753.32
4. The dress Sally wants cost $85.15. If the price was reduced by $12.78, how much will she pay?
   $72.37
5. Melissa went to the mall and noticed that the price of a coat she wanted was cut in half! The original price was $58.22. What is the sales price?
   $29.11
6. Tyler decided that he wanted a dog. He went to the pet store and bought one for $42.95. Tyler also bought three bags of food for $12.55 a bag. How much did Tyler spend altogether?
   $80.60
7. Christopher decided to make his grandmother a birdhouse instead of buying her one. The materials for the birdhouse totaled $21.99. the cost of a new birdhouse is $37.23. How much did Christopher save?
   $15.24
8. Jim thinks that snow skiing looks like lots of fun. He decided he wants to try it. First he needs equipment. He bought a pair of skis for $129.78, a pair of boots for $62.22, poles for $12.95, a hat for $2.50, a coat for $49.95, ski pants for $27.50 and gloves for $11.25. How much did Jim spend altogether?
   $296.15

40

1. $0.125 = \frac{1}{8}$
2. $0.\overline{6} = \frac{2}{3}$
3. $0.36 = \frac{9}{25}$
4. $0.\overline{46} = \frac{46}{99}$
5. $0.6875 = \frac{11}{16}$
6. $0.91\overline{6} = \frac{11}{12}$
7. $0.625 = \frac{5}{8}$
8. $0.\overline{27} = \frac{3}{11}$
9. $0.3\overline{8} = \frac{7}{18}$
10. $0.55 = \frac{11}{20}$
11. $0.5625 = \frac{9}{16}$
12. $0.775 = \frac{31}{40}$

41

.3	$\frac{1}{20}$	2.1
3.1	2.8	$\frac{8}{25}$
4	.1	$\frac{1}{2}$

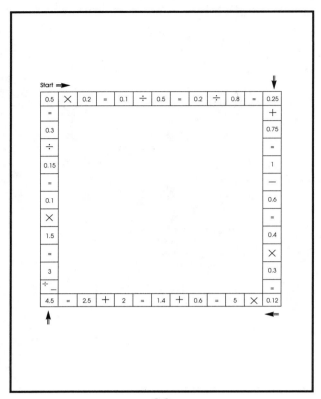

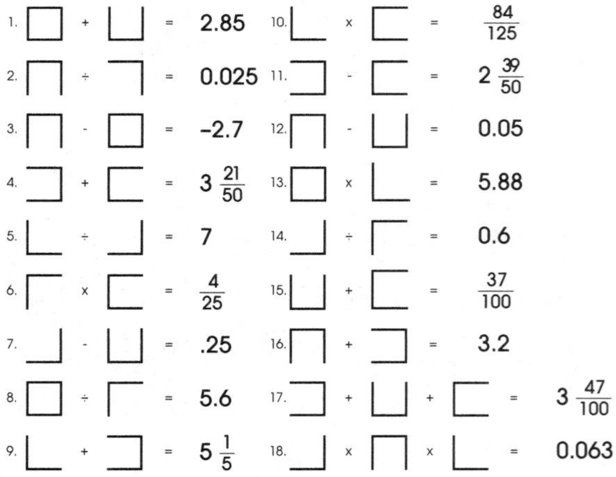

42

© Carson-Dellosa • CD-704384

Answer Key

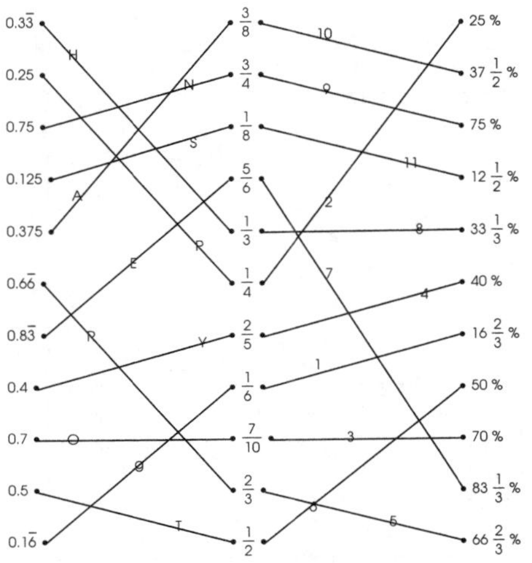

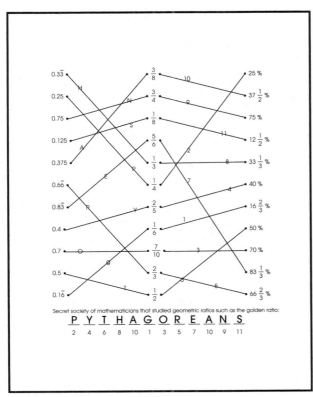

Secret society of mathematicians that studied geometric ratios such as the golden ratio:

P Y T H A G O R E A N S

2 4 6 8 10 1 3 5 7 10 9 11

43

1. What did Amelia Earhart's father say the first time he saw her fly an air plane?
   0.115 x 3 + 10141 x 5 = 50705.345
   Flip diᴉʇ She solos
2. What did Farmer Macgregor throw at Peter Rabbit to chase him out of the garden?
   (27 x 109 + 4 – 0.027) 2 x 9 = 53045.514
   Flip diᴉʇ his shoes
3. What did Snoopy add to his doghouse as a result of his dogfights with the Red Baron?
   7 (3 x 303 + 50) x 8 = 53704
   Flip diᴉʇ holes
4. What kind of double does a golfer want to avoid at the end of a round of golf?
   4 (1956 x 4 +153) = 31908
   Flip diᴉʇ BOGIE
5. What did the little girl say when she was frightened by the ghost?
   0.07 x 0.111 x 5 + 0.00123 = 0.04008
   Flip diᴉʇ BOO HOO

44

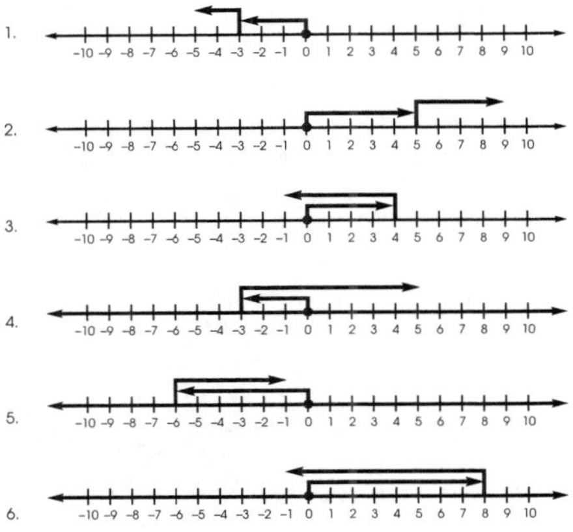

1. $^{-}5$
2. 9
3. $^{-}1$
4. 5
5. $^{-}1$
6. $^{-}1$

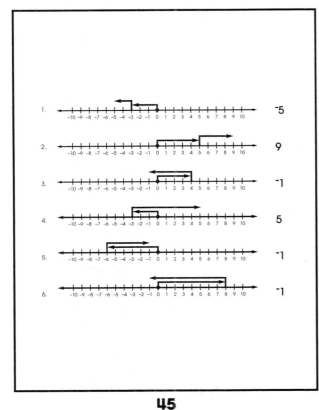

45

1. 6 + 8 = 14
2. $^{-}9 + {}^{-}23 =$ $^{-}32$
3. 25 + 37 = 62
4. $^{-}85 + {}^{-}19 =$ $^{-}104$
5. 132 + 899 = 1031
6. $^{-}104 + {}^{-}597 =$ $^{-}701$
7. $^{-}642 + {}^{-}33 =$ $^{-}675$
8. 88 + 298 = 386
9. $^{-}45 + {-}68 =$ $^{-}113$
10. $^{-}12 + {}^{-}18 + {}^{-}35 =$ $^{-}65$
11. 21 + 108 +111 = 240
12. $^{-}62 + {}^{-}33 + {}^{-}12 =$ $^{-}107$
13. 17 + 39 + 44 = 100
14. $^{-}18 + {}^{-}18 + {}^{-}18 =$ $^{-}54$
15. 19 + 42 + 647 = 708
16. $^{-}29 + {}^{-}108 + {}^{-}337 + {}^{-}503 =$ $^{-}977$

46

Answer Key

1. 21 + -87 = -66
2. -63 + 59 = -4
3. 12 + -12 = 0
4. -28 + 82 = 54
5. -32 + 97 = 65
6. -53 + 74 = 21
7. 132 + -87 = 45
8. 212 + -99 = 113
9. -331 + 155 = -176
10. -413 + 521 = 108
11. 8,129 + -6,312 = 1,817
12. -11,332 + 566 = -10,766
13. 1,627 + -7,193 = -5,566
14. 7,864 + -6,329 = 1,535
15. -10,822 + 6,635 = -4,187
16. 13,894 + -81,139 = -67,245
17. -16,742 + 65,524 = 48,782
18. -56,814 + 73,322 = 16,508
19. 101,811 + -322,885 = -221.074
20. 562,493 + -112,819 = 449,674
21. 116,667 + -912,182 = -795,515
22. -629,922 + 81,962 = -547,960
23. -196,322 + 422,899 = 226,577
24. 467,833 + -36,838 = 430,995

47

3	-1	5	9	-3	-7	0	4	-8	2
2	6	0	-4	-8	2	-7	1	5	-3
-9	4	-8	1	4	7	0	-3	6	2
4	-8	1	-5	9	-6	2	-6	0	-9
-3	8	2	-6	-3	7	-1	-5	9	8
5	-8	1	-4	7	-1	-5	9	-2	2
-6	0	-7	3	-7	1	5	9	2	0
3	-7	4	-8	2	6	0	4	-9	-5
5	8	-2	6	0	-3	6	-9	-2	9
4	2	-4	-8	1	6	0	4	1	6

48

1. 19 – 23 = -4
2. -8 – 7 = -15
3. 35 – 20 = 15
4. -46 – -18 = -28
5. -118 – 12 = -130
6. 7 – -103 = 110
7. 211 – 108 = 103
8. -9 – -16 = 7
9. 63 – 72 = -9
10. -93 – 117 = -210
11. 45 – -50 = 95
12. -18 – -12 = -6
13. 21 – 82 = -61
14. -831 – 616 = -1,447
15. -632 – -714 = 82
16. 1,192 – -983 = 2175

49

1. 7 – 13 = -6
2. -17 – 9 = -26
3. -11 – 7 = -18
4. -24 – -23 = -1
5. 2 – 25 = -23
6. 0 – -14 = 14
7. -3 – -7 = 4
8. -8 – -27 = 19
9. -29 – 36 = -65
10. -72 – -84 = 12
11. 63 – 94 = -31
12. 77 – -27 = 104
13. -23 – -96 = 73
14. -70 – 18 = -88
15. 318 – -864 = 1,182
16. -626 – 118 = -744
17. 553 – -764 = 1,317
18. -832 – 1,129 = -1,961
19. 6,793 – -8,329 = 15,122
20. -7,624 – 11,652 = -19,276
21. 108,719 – -96,989 = 205,708
22. -832,629 – -163,864 = -668,765
23. -629,299 – 532,106 = -1,161,405
24. 735,300 – -800,919 = 1,536,219

50

© Carson-Dellosa • CD-704384

Answer Key

1. -6 + -8 =	-14	13. -23 − -28 =	5
2. -10 − 3 =	-13	14. 0 − 31 =	-31
3. -14 + 20 =	6	15. -40 − 35 =	-75
4. 31 − -9 =	40	16. 73 + -19 =	54
5. -17 + 9 =	-8	17. -231 − -231 =	0
6. -8 − -27 =	19	18. -107 + -293 =	-400
7. -33 − 36 =	-69	19. 52 + -41 − 60 =	-49
8. 19 + -32 =	-13	20. -85 − -106 + 18 =	39
9. 112 − -52 =	164	21. 81 − 165 − -75 =	-9
10. 8 − -7 =	15	22. -16 + 312 + -621 =	-325
11. 24 + -24 =	0	23. -121 + -632 − -11 =	-742
12. 508 − 678 =	-170	24. -553 − -632 + -85 =	-6

51

1. (-3) (-6) =	18	11. (31) (-31) (31) =	-29,791
2. (14) (-4) =	-56	12. (-4) (-18) (28) =	2016
3. (25) (2) =	50	13. (-53) (-14) (-7) =	-5194
4. (20) (-49) =	-980	14. (32) (125) (11) =	44,000
5. (75) (15) =	1125	15. (-37) (-18) (-5) (2) =	-6,660
6. (-30) (-30) =	900	16. (111) (-63) (19) =	-132,867
7. (-17) (23) =	-391	17. (20) (-7) (35) (-3) =	14,700
8. (-218) (-32) =	6976	18. (16) (-8) (-10) (-1) =	-1,280
9. (801) (-37) =	-29,673	19. (-9) (-29) (32) (2) =	16,704
10. (-89) (-321) =	28,569	20. (-18) (-6) (-21) (-30) =	68,040

52

1. -4 • 15 =	-60	10. (1.2) (-5) =	-6
2. (-6) (-8) =	48	11. (6.5) (-1) (-3) =	19.5
3. (-10) (-3) (4) =	120	12. $(-\frac{5}{8})(-\frac{2}{3}) =$	$\frac{5}{12}$
4. (-21) (-4) (0) =	0	13. $(\frac{3}{8})(\frac{5}{6}) =$	$\frac{5}{16}$
5. (-3) (-3) (-3) =	-27	14. $(-12)(-\frac{1}{3})(\frac{3}{4}) =$	3
6. 14 (-6) =	-84	15. $(-6\frac{2}{3})(3\frac{3}{4}) =$	-25
7. -40 x -9 =	360	16. $(-2)^3 =$	-8
8. (4) (-2) (3) (-1) (5) (-6) =	-720	17. $(-3)^2 =$	9
9. (-1) (-1) (-1) (-1) (-1) (-1) =	1	18. $(-1)^{99} =$	-1

Complete the statements with either a positive or negative.

19. A problem with an even number of negative factors will have a **positive** product.

20. A problem with an odd number of negative factors will have a **negative** product.

53

1. -49 ÷ 7 =	-7	9. $\frac{17}{-17} =$	-1
2. 100 ÷ -4 =	-25	10. $\frac{-72}{-18} =$	4
3. -75 ÷ -15 =	5	11. $\frac{-195}{13} =$	-15
4. -84 ÷ 21 =	-4	12. $\frac{-23}{-1} =$	23
5. -120 ÷ 5 =	-24	13. $\frac{200}{10} =$	20
6. 57 ÷ -19 =	-3	14. $\frac{270}{-45} =$	-6
7. -288 ÷ -4 =	72	15. $\frac{-343}{7} =$	-49
8. 804 ÷ 67 =	12	16. $\frac{-1125}{-45} =$	25

54

Answer Key

1. $^-91 \div 7 =$ $^-13$
2. $36 \div (^-9) =$ $^-4$
3. $^-54 \div (^-9) =$ 6
4. $75 \div 15 =$ 5
5. $0 \div (^-7) =$ 0
6. $\frac{56}{^-7} =$ $^-8$
7. $\frac{^-72}{^-12} =$ 6
8. $\frac{102}{^-17} =$ $^-6$
9. $600 \div 24 =$ 25
10. $\frac{144}{^-12} =$ $^-12$
11. $^-48 \div 3 =$ $^-16$
12. $^-1.5 \div (^-3) =$ 0.5
13. $2.4 \div (^-1.2) =$ $^-2$
14. $^-1.44 \div (.3) =$ $^-4.8$
15. $\frac{0}{^-4.12} =$ 0
16. $\frac{1}{8} \div -\frac{6}{5} =$ $-\frac{5}{48}$
17. $-\frac{3}{7} \div -\frac{8}{21} =$ $1\frac{1}{8}$
18. $^-10 \div \frac{1}{3} =$ $^-30$
19. $-\frac{3}{4} \div (^-12) =$ $\frac{1}{16}$
20. $^-15 \div \frac{3}{5} =$ $^-25$
21. $\frac{4}{5} \div (-\frac{3}{10}) =$ $^-2\frac{2}{3}$
22. $-\frac{3}{8} \div (-\frac{3}{4}) =$ $\frac{1}{2}$
23. $\frac{5}{6} \div \frac{4}{9} =$ $1\frac{7}{8}$
24. $^-6\frac{2}{3} \div 3\frac{3}{4} =$ $^-1\frac{7}{9}$

55

A. $^-81 \div {^-9} =$ 9

B. $13 \div {^-13} =$ $^-1$

C. $^-60 \div 10 =$ $^-6$

D. $^-88 \div {^-11} =$ 8

E. $^-104 \div 8 =$ -13

F. $^-147 \div {^-21} =$ 7

G. $80 \div {^-5} =$ $^-16$

H. $52 \div 4 =$ 13

I. $^-150 \div {^-6} =$ 25

J. $\frac{^-102}{17} =$ $^-6$

K. $\frac{^-75}{^-5} =$ 15

L. $\frac{196}{^-14} =$ $^-14$

M. $\frac{1378}{^-26} =$ $^-53$

N. $\frac{^-468}{^-26} =$ 18

O. $\frac{253}{11} =$ 23

P. $\frac{^-465}{^-31} =$ 15

Q. $\frac{^-552}{^-23} =$ 24

R. $\frac{^-1824}{^-48} =$ 38

William I of Normandy conquered England in → → → ↓
(A+B+C+D+E+F) – (G+H) – (I÷(J+K+L)) – M•N + (O+P+Q+R) = 1066

(_ + _ + _ + _ + _ + _) – (_ + _) – (_ ÷ (_ + _ + _)) – _ • _ + (_ + _ + _ + _)

$(9 + {^-1} + {^-6} + 8 + {^-13} + 7) - (^-16 + 13) - (25 \div (^-6 + 15 + {^-14})) - {^-53} \cdot 18 + (23 + 15 + 24 + 38)$

56

1. $^-41 + {^-125} =$ $^-166$
2. $79 - 88 =$ $^-9$
3. $^-3 \bullet {^-4} =$ 12
4. $\frac{^-125}{5} =$ $^-25$
5. $19 \bullet {^-24} =$ $^-456$
6. $\frac{^-123}{41} =$ $^-3$
7. $82 + {^-95} =$ $^-13$
8. $27 - {^-46} =$ 73
9. $^-31 - {^-32} =$ 1
10. $\frac{^-825}{^-33} =$ 25
11. $^-34 + 52 + {^-18} =$ 0
12. $14 \bullet {^-12} \bullet 3 =$ $^-504$
13. $\frac{^-185}{5} \bullet -4 =$ 148
14. $76 - 19 + {^-60} =$ $^-3$
15. $17 - {^-12} - 22 =$ 7
16. $100 \bullet {^-4} \bullet 40 =$ $^-16{,}000$
17. $\frac{54}{^-9} + \frac{33}{11} + \frac{24}{8} =$ 0
18. $^-51 \div 17 =$ $^-3$
19. $4 - 8 + {^-9} =$ $^-13$
20. $-\frac{98}{49} \bullet {^-10} =$ 20
21. $(256 \div {^-16}) \bullet {^-3} =$ 48
22. $(^-18 - {^-26} + {^-13}) \bullet {^-2} =$ 10
23. $(202 + {^-196} - 321) \div {^-5} =$ 63
24. $(\frac{^-575}{23} - 18) \bullet {^-11} =$ 473

57

1. An elevator started at the first floor and went up 18 floors. It then cam, down 11 floors and went back up 16. At what floor was it stopped?

   23

2. At midnight, the temperature was 30° F. By 6:00 a.m., it had dropped 5° and by noon, it had increased by 11°. What was the temperature at noon?

   36°

3. Some number added to 5 is equal to $^-11$. Find the number.

   $^-16$

4. From the top of a mountain to the floor of the valley below is 4,392 feet. If the valley is 93 feet below sea level, what is the height of the mountain?

   4299 feet

5. During one week, the stock market did the following: Monday rose 18 points, Tuesday rose 31 points, Wednesday dropped 5 points, Thursday rose 27 points and Friday dropped 38 points. If it started out at 1,196 on Monday, what did it end up on Friday?

   1,229

6. An airplane started at 0 feet. It rose 21,000 feet at takeoff. It then descended 4,329 feet because of clouds. An oncoming plane was approaching, so it rose 6.333 feet. After the oncoming plane passed, it descended 8,453 feet, at what altitude was the plane flying?

   14,551

7. Some number added to $^-11$ is 37. Divide this number by $^-12$. Then, multiply by $^-8$. What is the final number?

   32

8. Jim decided to go for a drive in his car. He started out at 0 miles per hour (mph). He then accelerated 20 mph down his street. Then, to get on the highway he accelerated another 35 miles per hour. A car was going slow in front of him so he slowed down 11 mph. He then got off the highway, so he slowed down another 7 mph. At what speed is he driving?

   37 mph

58

© Carson-Dellosa • CD-704384

Answer Key

1. $^{-}1.6 + 1\frac{7}{10} =$ **0.1**
   (Hint: $1\frac{7}{10} = 1.7$)
2. $0 - 6\frac{1}{2} + {^{-}3} =$ **$^{-}9\frac{1}{2}$**
3. $\frac{^{-}3}{4} + 5 - \frac{1}{2} =$ **$3\frac{3}{4}$**
4. $9 - 10.2 + {^{-}8.6} =$ **$^{-}9.8$**
5. $\frac{1}{2} + 1\frac{1}{2} - 1\frac{1}{3} =$ **$\frac{2}{3}$**
6. $6.75 - 3\frac{1}{2} + 2.55 =$ **5.8**
   (Hint: $3\frac{5}{10} = 3.5$)
7. $3\frac{3}{7} - {^{-}1\frac{1}{7}} + \frac{3}{7} =$ **5**
8. $^{-}7 - {^{-}2\frac{3}{4}} + {^{-}5\frac{1}{4}} =$ **$^{-}9\frac{1}{2}$**
9. $7\frac{1}{10} + {^{-}7.25} - 11.39 =$ **$^{-}11.54$**
10. $^{-}8\frac{1}{4} + {^{-}3\frac{3}{12}} - 7\frac{2}{3} =$ **$^{-}19\frac{1}{6}$**
11. $^{-}5 - 7\frac{1}{8} + {^{-}3\frac{5}{12}} =$ **$^{-}15\frac{13}{24}$**
12. $3\frac{3}{10} + {^{-}3.38} - 6\frac{6}{10} =$ **$^{-}6.68$**

59

1. $^{-}3\frac{5}{10} + 8 =$ **4.5**
2. $^{-}5\frac{3}{7} + {^{-}3\frac{3}{14}} =$ **$^{-}8\frac{9}{14}$**
3. $6\frac{1}{6} - 6\frac{3}{10} =$ **$-\frac{2}{15}$**
4. $^{-}8 + 15.32 =$ **7.32**
5. $^{-}8\frac{3}{10} - {^{-}5.9} =$ **$^{-}2.4$**
6. $13 - 5\frac{3}{5} =$ **$7\frac{2}{5}$**
7. $12\frac{1}{9} + {^{-}5\frac{2}{3}} =$ **$6\frac{4}{9}$**
8. $^{-}11.03 - {^{-}21.6} =$ **10.57**
9. $^{-}7\frac{3}{10} - 16.53 =$ **$^{-}23.83$**
10. $31\frac{8}{9} + {^{-}27\frac{27}{81}} =$ **$4\frac{5}{9}$**
11. $11 - 18.6 + {^{-}3\frac{3}{10}} =$ **$^{-}10.9$**
12. $^{-}5\frac{2}{10} + 16.7 - 3\frac{1}{5} =$ **8.3**
13. $13\frac{1}{3} + {^{-}12} + {^{-}7\frac{7}{12}} =$ **$^{-}6\frac{1}{4}$**
14. $41.32 + {^{-}18.7} - 16.21 =$ **6.41**
15. $^{-}18.75 - 5\frac{3}{4} - 7\frac{5}{12} =$ **$^{-}31\frac{11}{12}$**
16. $^{-}15 - 21\frac{1}{7} + 18\frac{2}{49} =$ **$^{-}18\frac{5}{49}$**
17. $7\frac{2}{3} + {^{-}8\frac{4}{9}} - {^{-}16\frac{1}{6}} =$ **$15\frac{7}{18}$**
18. $^{-}31.5 - {^{-}3\frac{7}{10}} + 21 =$ **$^{-}6.8$**
19. $25\frac{1}{5} - 17.3 + {^{-}11\frac{2}{11}} =$ **$^{-}3\frac{31}{110}$**
20. $19.25 - {^{-}6\frac{3}{4}} + 12\frac{5}{12} =$ **$38\frac{5}{12}$**

60

1. $^{-}1\frac{2}{3} \bullet {^{-}3\frac{1}{5}} =$ **$5\frac{1}{3}$**
2. $4\frac{5}{9} \div -\frac{10}{27} =$ **$^{-}12.3$**
3. $4\frac{1}{4} \bullet 3\frac{1}{5} =$ **$13\frac{3}{5}$**
4. $^{-}9\frac{3}{8} \div {^{-}3\frac{9}{12}} =$ **2.5**
5. $-\frac{3}{8} \bullet 4 \bullet \frac{4}{9} =$ **$-\frac{2}{3}$**
6. $^{-}9\frac{3}{5} \div \frac{12}{5} \bullet {^{-}4} =$ **16**
7. $^{-}4.1 \bullet {^{-}5.2} \div 4 =$ **5.33**
8. $6.2 \bullet 3 \bullet -\frac{1}{2} =$ **$^{-}9.3$**
9. $({^{-}2\frac{1}{2}})({^{-}2\frac{1}{2}}) \div 0.5 =$ **12.5**
10. $-\frac{6}{7} \bullet -\frac{5}{12} \bullet -\frac{2}{15} =$ **$-\frac{1}{21}$**
11. $5\frac{2}{3} \bullet 9.81 \bullet 0 =$ **0**
12. $12 \bullet 3\frac{1}{4} \bullet {^{-}2\frac{2}{3}} =$ **$^{-}104$**

61

1. $^{-}9\frac{3}{5} \bullet \frac{5}{12} =$ **$^{-}4$**
2. $-\frac{16}{7} \div \frac{12}{35} =$ **$^{-}6\frac{2}{3}$**
3. $4\frac{1}{2} \bullet {^{-}2\frac{2}{7}} =$ **$^{-}10\frac{2}{7}$**
4. $^{-}5\frac{5}{6} \div 2\frac{1}{3} =$ **$^{-}2\frac{1}{2}$**
5. $^{-}8\frac{1}{3} \bullet {^{-}2\frac{2}{5}} =$ **20**
6. $16\frac{1}{8} \div 14\frac{1}{3} =$ **$1\frac{1}{8}$**
7. $^{-}37.6 \bullet 0.03 =$ **$^{-}1.128$**
8. $^{-}16.188 \div {^{-}4.26} =$ **3.8**
9. $^{-}1.75 \bullet {^{-}3.4} =$ **5.95**
10. $^{-}3.45 \div 1\frac{1}{2} =$ **$^{-}2.3$**
11. $^{-}8 \div {^{-}1\frac{1}{3}} \bullet {^{-}5} =$ **$^{-}30$**
12. $4.498 \div {^{-}1.73} \bullet {^{-}1.2} =$ **3.12**
13. $-\frac{5}{7} \div -\frac{1}{14} \bullet -\frac{1}{2} =$ **$^{-}5$**
14. $^{-}6\frac{2}{3} \bullet 2.75 \div {^{-}1\frac{2}{3}} =$ **11**
15. $-\frac{3}{8} \div {^{-}3} \bullet \frac{4}{5} =$ **$\frac{1}{10}$**
16. $12\frac{3}{8} \bullet {^{-}2\frac{2}{3}} \div 2.5 =$ **$^{-}13.2$**
17. $-\frac{5}{6} \bullet 4\frac{1}{4} \bullet -\frac{3}{5} =$ **$2\frac{1}{8}$**
18. $^{-}3\frac{1}{5} \div 4\frac{2}{5} \div {^{-}1\frac{1}{7}} =$ **$\frac{7}{11}$**
19. $3\frac{3}{5} \bullet {^{-}1.46} =$ **$^{-}5.256$**
20. $4\frac{2}{3} \div -\frac{6}{7} \bullet \frac{9}{10} =$ **$^{-}4.9$**

62

1. $^{-}28 \div 7 + 2\frac{1}{3} =$ $^{-}1\frac{2}{3}$
2. $\frac{1}{2}(^{-}16 - 4) =$ $^{-}10$
3. $^{-}9 \div {}^{-}3 + 4 \bullet -\frac{1}{4} - 20 \div 5 =$ $^{-}2$
4. $\frac{1}{3}((^{-}18 + 3) + (5 + 7) \div {}^{-}4) =$ $^{-}6$
5. $(8\frac{1}{3} + 3\frac{2}{3}) \div 4 - {}^{-}16 =$ 19
6. $\frac{(80 \bullet \frac{1}{2}) + 35}{^{-}10 + 25} =$ 5
7. $2\,(^{-}6\,(3 - 12) - 17) =$ 74
8. $\frac{1}{4}(20 + 72 \div {}^{-}9) =$ 3
9. $3 \bullet 2\,(4 + (9 \div 3)) =$ 42
10. $50 \div ((4 \bullet 5) - (36 \div 2)) + {}^{-}91 =$ $^{-}66$

63

Problem	Solution	Clue	Word
1. 501 x 7	3507	To not win	LOSE
2. $10^3 - 3 \times 131$	607	Type of cabin	LOG
3. $17^2 + 7^2$	338	It buzzes.	BEE
4. 67,077 ÷ 87	771	Sick	ILL
5. 2 • (2 • 1900 + 3 • 23)	7738	It rings.	BELL
6. $2^9 + 2$	514	Not hers	HIS
7. $279^2 - (500 - 4)$	77345	Nautilus ______	SHELL
8. $3^3 \times 100 + 3 \times 115$	3045	Worn on foot	SHOE
9. $22{,}416 \div 2^2$	5604	Big pigs	HOGS
10. 473,720 – 12,345	461375	Snow vehicle	SLEIGH
11. 3 x 5 x 246 +15	3705	Bottom of shoe	SOLE
12. 4,738 – 1,234	3504	Fire equipment	HOSE
13. $60^2 + 4 \times 26$	3704	Center of a donut	HOLE
14. 11 x (60 – 2)	638	To plead	BEG
15. 5787 ÷ 9 x 12	7716	Fish organ	GILL
16. 12,345 + 23,456 – 465	35336	They "honk".	GEESE
17. 8 x 100 + 8 – 1	807	Tennis shot	LOB
18. $50 \times 700 + 3 \times 6^2$	35108	Capital of Idaho	BOISE
19. 50 x 110 + (10 – 3)	5507	Not a win	LOSS
20. $64{,}118 - 80^2$	57718	Ducks' beaks	BILLS

A googol is 10^{100} or 1 followed by 100 zeros.
What number would result in the "calculator word" googol? 706006

64

1. 2.5 **=** $2\frac{17}{34}$
2. 1.049 **<** 1.49
3. $^{-}0.\overline{3}$ **<** $^{-}0.3$
4. 15.62 **>** 1.562
5. 8156.6 **<** 8166.6
6. $^{-}7\frac{4}{5}$ **=** $^{-}7\frac{24}{30}$
7. $^{-}8\frac{7}{8}$ **<** $^{-}8.857$
8. 329.93 **>** 32.993
9. 982.61 **<** 7662.8
10. $13\frac{5}{8}$ **>** 13.6

1. 6.41, 6.411, 6.4111 — **3 2 1**
2. $^{-}2\frac{9}{14}, {}^{-}2\frac{5}{8}, {}^{-}2\frac{4}{7}$ — **3 2 1**
3. $11.6, 11\frac{2}{3}, 11\frac{14}{25}$ — **2 1 3**
4. $^{-}0.030, -\frac{33}{100}, {}^{-}0.003$ — **2 3 1**
5. $7\frac{5}{8}, 7\frac{3}{4}, 7.775$ — **3 2 1**
6. $^{-}10\frac{3}{4}, {}^{-}10.82, {}^{-}10\frac{2}{3}$ — **2 3 1**
7. $3.08, 3\frac{4}{5}, 3\frac{3}{5}$ — **3 1 2**
8. $^{-}1.35, {}^{-}1\frac{1}{8}, {}^{-}1\frac{1}{4}$ — **3 1 2**

65

1. $|{}^{-}12| =$ 12
2. $-|5\frac{1}{2}| =$ $^{-}5\frac{1}{2}$
3. $|{}^{-}5| + |9| =$ 14
4. $7 + |{}^{-}3| =$ 10
5. $|7| + |{}^{-}7| =$ 14
6. $-(^{-}3 + 4) =$ $^{-}1$
7. $-(9 - 9) =$ 0
8. $|9| - |{}^{-}12| =$ $^{-}3$
9. $|{}^{-}3| + |9| - 6 =$ 6
10. $^{-}3\,|5| - |5| =$ $^{-}20$
11. $|{}^{-}25| - |{}^{-}14| =$ 11
12. $^{-}18 + (-(^{-}13)) =$ $^{-}5$
13. $|1 - 3| + 5 =$ 7
14. $\frac{-|{}^{-}3 + 5|}{-9 + (^{-}(^{-}1))} =$ $\frac{1}{4}$
15. $-(2n - (^{-}7)) =$ $^{-}2n - 7$
16. $-(^{-}2x + {}^{-}3y) =$ $2x + 3y$
17. $-(6x - 4y) =$ $^{-}6x + 4y$
18. $10m - (^{-}2n) =$ $10m + 2n$
19. $^{-}2\,(3m^2 - 2m - 1) =$ $^{-}6m^2 + 4m + 2$
20. $^{-}3\,(4x - 6y) =$ $^{-}12x + 18y$

66

© Carson-Dellosa • CD-704384

Answer Key

1. 1 – 3 equals
   a. ˉ4 — **b. ˉ2** (circled) — c. 4 — d. 2

2. If x and y are positive integers and if $\frac{x}{y} = 1$ and $(x + y)^2 = z$, which of the following can equal z?
   a. 5 — b. 9 — **c. 16** (circled) — d. 25

3. (ˉ1) (ˉ2) (ˉ3) (+4) =
   a. ˉ10 — b. 24 — **c. ˉ24** (circled) — d. ˉ36

4. (ˉ2) – (ˉ5) =
   a. ˉ7 — b. ˉ3 — **c. 3** (circled) — d. 7

5. (ˉ5) + (ˉ2) =
   a. ˉ7 (circled) — b. ˉ3 — c. 3 — d. 7

6. $(\frac{1}{2}) \div (-\frac{7}{8}) =$
   a. $-\frac{4}{7}$ (circled) — b. $-\frac{7}{16}$ — c. ˉ$1\frac{3}{4}$ — d. ˉ$2\frac{2}{7}$

7. 7 – ((ˉ8) + (ˉ2))
   a. ˉ3 — b. 4 — c. 13 — **d. 17** (circled)

8. $\left| \frac{(^-18) + (^-2)}{(7) + (^-2)} \right|$
   a. $2\frac{2}{9}$ — **b. 4** (circled) — c. $3\frac{1}{5}$ — d. ˉ4

9. The integers ˉ2, ˉ7, 5, and ˉ5 written from least to greatest are
   a. ˉ2, ˉ5, ˉ7, 5 — b. ˉ5, ˉ7, ˉ2, 5 — **c. ˉ7, ˉ5, ˉ2, 5** (circled) — d. ˉ7, ˉ2, ˉ5, 5

10. Which of the following conditions will make $x - y$ a negative number?
    a. $y > x$ (circled) — b. $x > y$ — c. $y > 0$ — d. $x = y$

67

"Ancient History"

Start at (0, ˉ1)

(1, ˉ1)	(0, 3)
(1, ˉ3)	(ˉ1, 4)
(3, ˉ3)	(ˉ2, 3)
(3, ˉ1)	(ˉ3, 4)
(5, 0)	(ˉ4, 3)
(8, 0)	(ˉ5, 1)
(7, 1)	(ˉ8, 2)
(9, 0)	(ˉ5, 0)
(8, 2)	(ˉ3, ˉ1)
(5, 1)	(ˉ3, ˉ3)
(4, 3)	(ˉ1, ˉ3)
(3, 4)	(ˉ1, ˉ1)
(2, 3)	(0, ˉ1)
(1, 4)	End

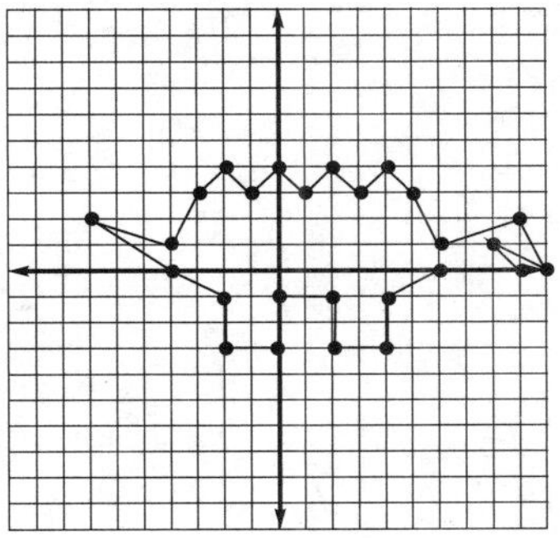

68

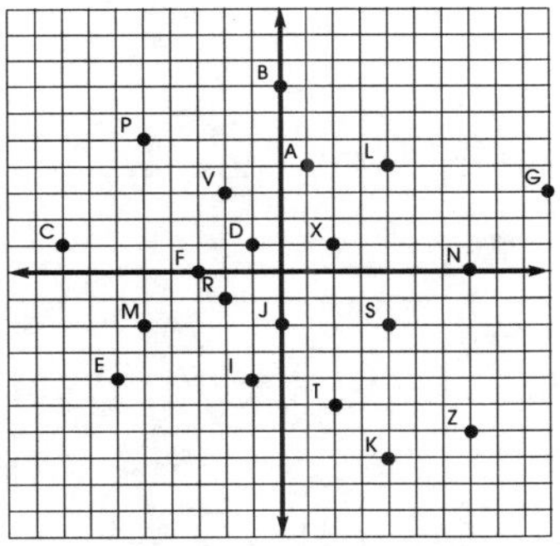

Find the coordinates associated with the following points.

1. A **(1, 4)**	6. C **(ˉ8, 1)**
2. K **(4, ˉ7)**	7. B **(0, 7)**
3. E **(ˉ6, ˉ4)**	8. S **(4, ˉ2)**
4. P **(ˉ5, 5)**	9. D **(ˉ1, 1)**
5. T **(2, ˉ5)**	10. N **(7, 0)**

Find the letter associated with each pair of coordinates.

11. (2, 1) **X**	16. (ˉ2, 3) **V**
12. (ˉ1, ˉ4) **I**	17. (ˉ3, 0) **F**
13. (10, 3) **G**	18. (4, 4) **L**
14. (7, ˉ6) **Z**	19. (ˉ5, ˉ2) **M**
15. (ˉ2, ˉ1) **R**	20. (0, ˉ2) **J**

69

I. Find the coordinates of the indicated point.

1. A (3, ˉ4)
2. I (ˉ2, ˉ6)
3. H (2, 9)
4. C (3, 0)
5. E (ˉ5, 1)
6. N (ˉ6, 4)

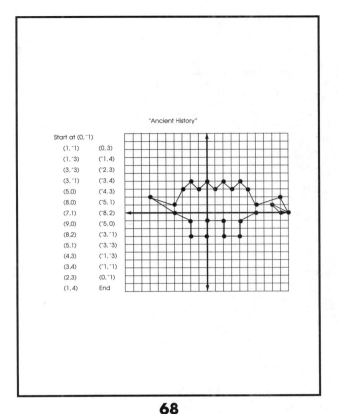

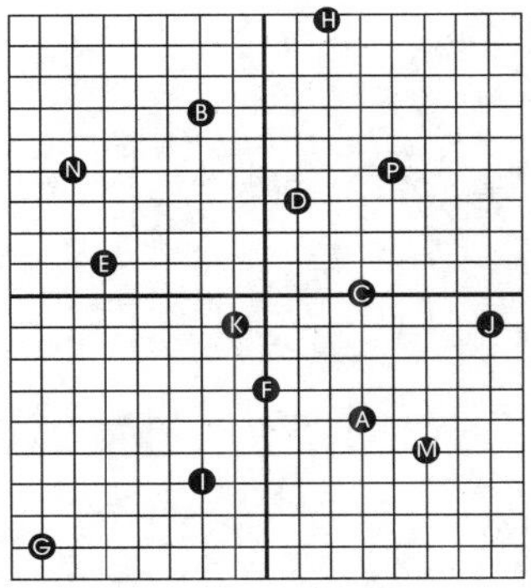

II. Name the graph (letter) of each ordered pair.

7. (ˉ2, 6) B
8. (0, ˉ3) F
9. (5, ˉ5) M
10. (ˉ1, ˉ1) K
11. (ˉ7, ˉ8) G
12. (7, ˉ1) J

13. The coordinates are equal.
    K and P

14. The y-coordinate is three times the x-coordinate.
    I and D

II. Name the quadrant or axis on which each point lies.

15. (ˉ4, 3) II
16. (0, 6) y - axis
17. (4, ˉ2) IV
18. (ˉ1, ˉ1) III
19. (ˉ2, 0) x - axis
20. (1, 2) I

70

Answer Key

Mike, Dale, Paul and Charlie are the athletic director, quarterback, pitc and goalie, but not necessarily in that order. From these five statements, identify the man in each position.

1. Mike and Dale were both at the ballpark when the rookie pitcher played his first game.

2. Both Paul and the athletic director had played on the same team in high school with the goalie.

3. The athletic director, who scouted Charlie, is planning to watch Mike during his next game.

4. Mike doesn't know Dale. One of these men is a quarterback.

	Quarterback	Goalie	Pitcher	Athletic Director
Mike	X			
Dale				X
Paul			X	
Charlie		X		

71

1. $a \bullet a \bullet a \bullet b$ — $a^3 b$
2. $mn \bullet mn \bullet mn \bullet mn$ — $(mn)^4$
3. $9 \bullet x \bullet x \bullet x \bullet x \bullet x \bullet y \bullet y \bullet z$ — $9x^5y^2z$
4. $5\,(c + 1)\,(c + 1)\,(c + 1)$ — $5\,(c + 1)^3$
5. $(a + b)$ squared — $(a + b)^2$
6. The quotient of 3 and the cube of $y + 2$ — $\frac{3}{(y + 2)^3}$
7. $x \bullet x \bullet y \bullet y \bullet y \bullet y \bullet z$ — x^2y^4z
8. $(-x)\,(-x)\,(-x)$ — $(-x)^3$
9. $3 \bullet ab \bullet ab \bullet ab \bullet ab$ — $3\,(ab)^4$ or $3a^4b^4$
10. The square of $x^2y - 3$ — $(x^2y - 3)^2$

1. x^5 — $^{-}1$
2. x^2yz — $^{-}6$
3. $4y^3z$ — $^{-}96$
4. $x^5y^4z^3$ — 432
5. $-(xyz)$ — $^{-}6$
6. $10z^5$ — $^{-}2{,}430$
7. x^2yz^2 — 18
8. $^{-}2xy^2$ — 8
9. $\frac{x^2z^2}{z}$ — $^{-}3$
10. $11x^2$ — 11

72

1. $9x + 4x =$ $13x$
2. $17x + x =$ $18x$
3. $m + (^{-}4m) =$ $^{-}3m$
4. $^{-}7x - 8x =$ $^{-}15x$
5. $14a - 19a =$ $^{-}5a$
6. $-a + 9a =$ $8a$
7. $6xy + 5xy =$ $11xy$
8. $^{-}9m - m =$ $^{-}10m$
9. $15a + (^{-}11a) =$ $4a$
10. $^{-}14x + 13x =$ $-x$
11. $5x^2y + 13x^2y =$ $18x^2y$
12. $21xy + (^{-}9xy) =$ $12xy$
13. $17x + 1 =$ $17x + 1$
14. $3.5y - 7.2y =$ $^{-}3.7y$
15. $^{-}4.7y - 2.3y =$ $^{-}7y$
16. $3a + 5c - 9a =$ $^{-}6a + 5c$
17. $2x - 9x + 7 =$ $^{-}7x + 7$
18. $7x - 8 - 11x =$ $^{-}4x - 8$
19. $3x - 3y - 9x + 7y =$ $^{-}6x + 4y$
20. $17x + 4 - 3x =$ $14x + 4$
21. $3x - 7y - 12y =$ $3x - 19y$
22. $11a - 13a + 15a =$ $13a$
23. $17x + 5a - 3x - 4a =$ $14x + a$
24. $6x + 9y + 2x - 8y + 5 =$ $8x + y + 5$
25. $3xy + 4xy + 5x^2y + 6xy^2 =$ $7xy + 5x^2y + 6xy^2$
26. $^{-}25y - 17y + 6xy - 3xy =$ $^{-}42y + 3xy$

73

1. A. $5t + 3r + 9t - 10r$ — $14t - 7r$; E. $r + t - 8r + 13t$ — $^{-}7r + 14t$; I. $-r + 4t + 10t + 8r$ — $7r + 14t$
2. D. $12x - 3y + x + 2y$ — $13x - y$; E. $3\,(4x - 3y) + x + 3y$ — $13x - 6y$; F. $4\,(4x - 2y) - 3x + 7y$ — $13x - y$
3. E. $4\,(y - 7x) - y$ — $3y - 28x$; I. $^{-}30x - (^{-}2x)$ — $^{-}28x$; O. $^{-}7\,(4x + y) + 7y$ — $^{-}28x$
4. U. $6\,(x - y) - 3\,(3x + y)$ — $^{-}3x - 9y$; V. $3\,(3x - y) - 6y$ — $9x - 9y$; W. $4x + y - 7x - 10y$ — $^{-}3x - 9y$
5. Q. $3\,(r - 1) - 4r + 5$ — $-r + 2$; X. $2\,(3 - 2r) - 4\,(2 - r)$ — $^{-}2$; Z. $-r + 7 + 3r - 9 - 2r$ — $^{-}2$
6. L. $8\,(x + y) + 3\,(x + y)$ — $11x + 11y$; M. $10\,(x + y) + x + y$ — $11x + 11y$; N. $9\,(x + y) - 2\,(x + y)$ — $7x + 7y$
7. A. $3\,(2b - a) - (2a - b)$ — $7b - 5a$; B. $3\,(a + 2b) - (b + 2a)$ — $a + 5b$; C. $2\,(a + 2b) - (a - b)$ — $a + 5b$
8. I. $^{-}5\,(a - b) - 2\,(a - b) + 8\,(a - b)$ — $a - b$; U. $6\,(a - b) - 4\,(a - b) + (a - b)$ — $3a - 3b$; O. $(a - b) - (a - b) + (a - b)$ — $a - b$
9. R. $3\,(x - y) - 2\,(y - x)$ — $5x - 5y$; S. $2\,(x - y) - 3\,(y - x)$ — $5x - 5y$; T. $3\,(y - x) - 2\,(x - y)$ — $^{-}5x + 5y$
10. L. $^{-}4\,(x + 2\,(5xy - x))$ — $4x - 40xy$; M. $^{-}4\,(x + 5\,(^{-}3xy + x))$ — $^{-}24x + 60xy$; N. $^{-}2\,(3x + 3\,(^{-}10xy + 3x))$ — $^{-}24x + 60xy$

Two expressions in each problem are

E	Q	U	I	V	A	L	E	N	T
2	5	8	1	4	7	10	3	6	9

74

© Carson-Dellosa • CD-704384

Answer Key

75

1. $\frac{18 + {}^{-}6}{2} = a$ **6**
2. ${}^{-}3 \bullet 4 - 6 = c$ **$^{-}18$**
3. $4.5 - 6.2 = p$ **$^{-}1.7$**
4. $\frac{{}^{-}3}{8} \bullet {}^{-}4 - 1 = q$ **$\frac{1}{2}$**
5. $\frac{{}^{-}15 + {}^{-}27}{3} = x$ **$^{-}14$**
6. ${}^{-}8.1 \bullet 4.2 + 16 = g$ **$^{-}18.02$**
7. $\frac{1}{3} \bullet {}^{-}15 + {}^{-}10 = r$ **$^{-}15$**
8. $1\frac{3}{5} \div \frac{16}{45} = d$ **4.5**
9. $5 \bullet 7.32 - 18.19 = n$ **18.41**
10. $\frac{3}{4} \bullet {}^{-}16 + 8.12 = z$ **$^{-}3.88$**
11. $\frac{{}^{-}40 + 15}{5} + 6 = b$ **1**
12. $-\frac{2}{5} \div \frac{4}{15} + {}^{-}2\frac{1}{2} = t$ **$^{-}4$**

76

1. $7 + x = 3\frac{1}{2}$, if $x = {}^{-}3\frac{1}{2}$
   true
2. $y + 15 \div 6 = {}^{-}1\frac{1}{2}$, if $y = {}^{-}3$
   false
3. $\frac{f}{13} + {}^{-}3 = 0$, if $f = 69$
   false
4. $2x - 5.45 = 0.97$, if $x = 3.21$
   true
5. $8\frac{1}{3} + a = 15\frac{8}{15}$, if $a = 7\frac{2}{5}$
   false
6. $8 + (z - 32) = {}^{-}10$, if $z = 16$
   false
7. $11.5 + c = 28\frac{1}{4}$, if $c = 16\frac{3}{4}$
   true
8. $y(5 + 11) + 8 = 41$, if $y = 2$
   false
9. $3g + 5.26 - 11.9 = 12.64$, if $g = {}^{-}3$
   false
10. $5 + -\frac{16}{k} = {}^{-}3$, if $k = 2$
    true
11. $7\frac{1}{9} \div w = \frac{1}{18}$, if $w = 2\frac{17}{32}$
    false
12. $\frac{3(2q - q)}{8} + 29 = 32$, if $q = 8$
    true
13. $\frac{16.8 - 91.6}{m}$ 37.4, if $m = 2$
    false
14. $11\frac{1}{4} - f = 5\frac{1}{16}$, if $f = 16\frac{5}{16}$
    false

77

1. $4(a - 1) =$ **$^{-}2$**
2. $4a - 3y =$ **8**
3. $4(x - 3y) =$ **40**
4. $x(a + 6) =$ **26**
5. $6a + {}^{-}12a =$ **$^{-}3$**
6. $7(x + -y) =$ **42**
7. $6a(8a + 4y) =$ **$^{-}12$**
8. $3x + 2(a - y) =$ **17**
9. $x(ax + ay) =$ **4**
10. $ay + y - 5ax =$ **$^{-}13$**
11. $xy(2a + 3x - 2) =$ **$^{-}88$**
12. $4x - (xy + 2) =$ **22**
13. $5y - 8a + 6xy - 7x =$ **$^{-}90$**
14. $10x(8a + {}^{-}4y) + {}^{-}3y =$ **486**
15. $6xy - 2x(4a - 8y) =$ **$^{-}192$**
16. $(2a - x)(2x - 6) =$ **$^{-}6$**

78

1. ${}^{-}7(a + b) =$ **$^{-}7a - 7b$**
2. $x(y - 4) =$ **$xy - 4x$**
3. $-\frac{2}{3}(c - 12) =$ **$-\frac{2}{3}c + 8$**
4. ${}^{-}8(\frac{t}{2} + 6) =$ **$^{-}4t + {}^{-}48$**
5. $y({}^{-}16 + 2x) =$ **$2xy - 16y$**
6. $3(2a - 8b) =$ **$6a - 24b$**
7. $2x(3y + {}^{-}6) =$ **$6xy - 12x$**
8. $7({}^{-}5x + 8z) =$ **$56z - 35x$**
9. ${}^{-}5y(6z - 10) =$ **$50y - 30yz$**
10. ${}^{-}3x({}^{-}7 + 8y) =$ **$21x - 24xy$**

1. $9y + 6y - 2 =$ **$15y - 2$**
2. $25x - x + 2y =$ **$24x + 2y$**
3. $4a + 8b + 11a - 10b =$ **$15a - 2b$**
4. $13xy + 18xy - 20xy =$ **$11xy$**
5. ${}^{-}2m + 16 - 13m =$ **$16 - 15m$**
6. $4a + 7 + 3a - 8 - 3a =$ **$4a - 1$**
7. $16x + {}^{-}18y + 10x - 7y =$ **$26x - 25y$**
8. $6c - 8ab + 9c - 10 =$ **$15c - 8ab - 10$**
9. $18ab + {}^{-}6a + {}^{-}7b + 26ab + {}^{-}7b =$ **$44ab + {}^{-}6a + {}^{-}14b$**
10. $5x - 3x + 2xy + 31x + {}^{-}18xy =$ **$33x - 16xy$**

1. $3(a + b) + 2b =$ $3a + 5b$
2. $5a + 2a(5 - b) =$ $15a - 2ab$
3. $8 - 3(6 - 6a) =$ $^-10 + 18a$
4. $4a + 6(a + 8) =$ $10a + 48$
5. $^-2a - 3(b - 4a) =$ $10a - 3b$
6. $8(6a + 7b) - 11(2b + 8a) =$ $^-40a + 34b$
7. $^-6(a + 5b) - 3(^-7b - a) =$ $^-3a - 9b$
8. $2(a - b) + 3(a - b) - 4(a - b) =$ $a - b$
9. $4a + {}^-7(a + 2) =$ $^-3a - 14$
10. $6(a + 2b) + 8a - 16b =$ $14a - 4b$
11. $3a + {}^-2(a + b) =$ $a - 2b$
12. $2(3a - 4b) - 6a =$ ^-8b
13. $^-5(2a - 3b) + 5(3b - 2a) =$ $^-20a + 30b$
14. $4(11a - 9b) - 7(6a) =$ $2a - 36b$
15. $^-3(4a - 5b) - (a - b) =$ $^-13a + 16b$

$10 + 18a$ YOU	~~$^-10 + 18a$ ARE~~	~~$a - 2b$ SEE~~	$2a + 36b$ CANNOT	~~$3a + 5b$ LEAD~~	$12a - 8b$ INSTRUCT
$14a + 4b$ AN	~~$14a - 4b$ A~~	$40a + 34b$ ELDERLY	~~$^-40a + 34b$ HORSE~~	$3a - 14$ CANINE	~~$2a - 36b$ TO~~
~~$10a + 48$ WATER~~	~~$^-3a - 14$ BUT~~	$12a - 8b$ ON	~~$^-13a + 16b$ YOU~~	~~$^-3a - 9b$ CANNOT~~	$15a - 2b$ FRESH
~~$10a - 3b$ MAKE~~	~~$a - b$ HIM~~	~~$15a - 2ab$ DOWN~~	~~$^-20a + 30b$ DRINK~~	$20a - 30b$ PROCEDURES	~~^-8b HOME~~

Write the familiar proverb.

You can't teach an old dog new tricks.

79

	Column A		Column B	
B	1. $y + 12 = 8$	$^-4$	A. $y - 12 = {}^-12$	0
E	2. $\frac{y}{6} = {}^-2$	$^-12$	B. $2y = {}^-8$	$^-4$
D	3. $^-7y = {}^-84$	12	C. $y - 1 = {}^-7$	$^-6$
G	4. $^-42 = y - 20$	$^-22$	D. $y - {}^-12 = 24$	12
A	5. $92 + y = 92$	0	E. $\frac{y}{^-4} = 3$	$^-12$
H	6. $9 = 54y$	$\frac{1}{6}$	F. $y + 2 = 11$	9
C	7. $^-12 = y - 6$	$^-6$	G. $y + 11 = {}^-11$	$^-22$
J	8. $^-1 = \frac{y}{20}$	$^-20$	H. $12y = 2$	$\frac{1}{6}$
F	9. $27 = 3y$	9	I. $\frac{y}{^-2} = {}^-10$	20
I	10. $^-5 + y = 15$	20	J. $^-15 = y + 5$	$^-20$

80

1. $3(x + 8) = {}^-6$ $^-10$
2. $75 = {}^-5(a + 5)$ $^-20$
3. $^-8(y - 6) = {}^-16$ 8
4. $20 = 4(\frac{t}{4} - 2)$ 28
5. $17(x - 2) = {}^-34$ 0
6. $63 = 9(2 - a)$ $^-5$
7. $6(2 - \frac{x}{6}) = 1$ 11
8. $^-36 = 6(y - 2)$ $^-4$
9. $^-7(r + 8) = {}^-14$ $^-6$
10. $3(m + 5) = 42$ 9
11. $^-54 = 3(2 + 5m)$ $^-4$
12. $^-3(x - 7) + 2 = 20$ 1

81

1. Five less than a number $x - 5$
2. Three times the sum of a number and twelve $3(y + 12)$
3. Ten more than the quotient of c and three $10 + \frac{c}{3}$
4. Two increased by six times a number $2 + 6x$
5. Two-thirds of a number minus eleven $\frac{2}{3}y - 11$
6. Twice the difference between c and four $2(c - 4)$
7. The product of nine and a number, decreased by seven $9x - 7$
8. Six times a number plus seven times the number $6n + 7n$
9. A number increased by twice the number $n + 2n$
10. One-fourth times a number increased by eleven $\frac{1}{4}x + 11$
11. The quotient of a number and three decreased by five $\frac{n}{3 - 5}$
12. Twelve times the sum of a number and five times the number $12(n + 5n)$

82

© Carson-Dellosa • CD-704384

Answer Key

1. A number decreased by 16 is $^-26$. Find the number.
   $^-10$
2. One-fourth of a number is $^-60$. Find the number.
   $^-240$
3. The product of negative eight and a number is 104. Find the number.
   $^-13$
4. Twice a number is 346. Find the number.
   173
5. A number increased by negative twenty-seven is 110. Find the number.
   137
6. Tim weighs five pounds more than Mitchell. Find Mitchell's weight if Tim weighs ninety-three pounds.
   88
7. The cost of five books is $71.00. What is the cost of each book?
   $14.20
8. The cost of a filter is $4.00. What is the cost of six filters?
   $24.00

83

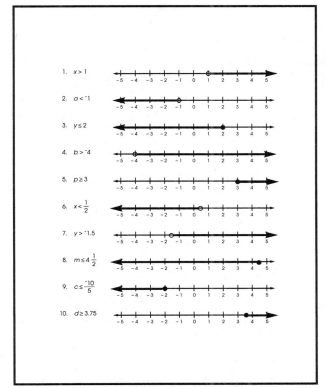

1. $x > 1$
2. $a < ^-1$
3. $y \leq 2$
4. $b > ^-4$
5. $p \geq 3$
6. $x < \frac{1}{2}$
7. $y > ^-1.5$
8. $m \leq 4\frac{1}{2}$
9. $c \leq \frac{^-10}{5}$
10. $d \geq 3.75$

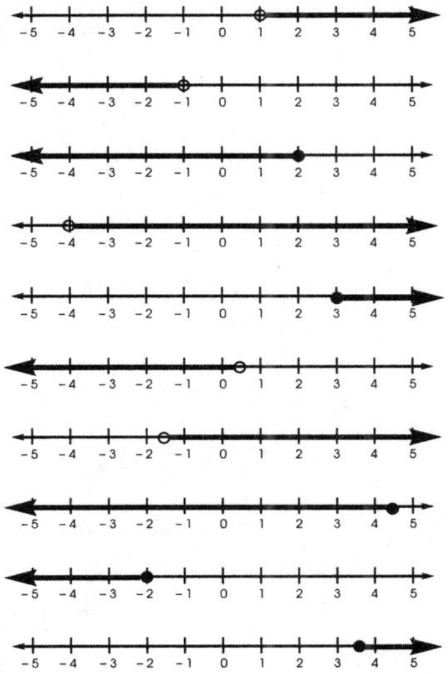

84

1. $g + 8 > 6$
   $g > ^-2$
2. $d - 7 > ^-3$
   $d > 4$
3. $^-3 > y + 1$
   $y < ^-4$
4. $a - 3 \leq 1$
   $a \leq 4$
5. $^-4 \leq 1 + c$
   $c \geq ^-5$
6. $x + \frac{1}{4} \geq 1\frac{1}{2}$
   $x \geq 1\frac{1}{4}$
7. $^-2.4 < n - 0.6$
   $n > ^-1.8$
8. $^-20 + m \leq ^-24$
   $m \leq 4$
9. $^-7.5 + x \geq ^-9$
   $x \geq ^-1.5$
10. $3\frac{1}{3} \leq \frac{2}{9} + c$
    $c \geq 3\frac{1}{9}$

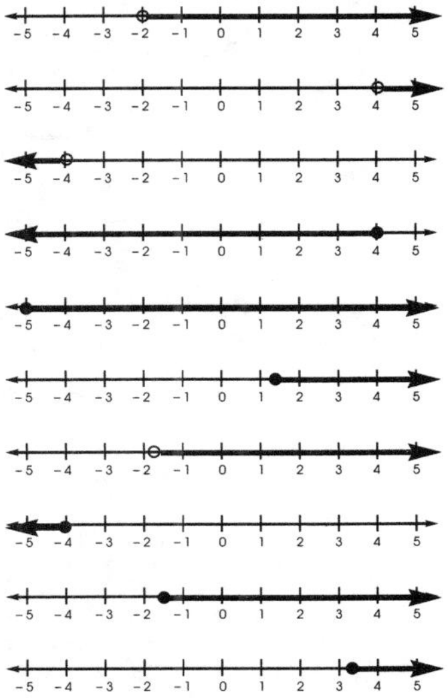

85

1. $11x > 22$ $x > 2$

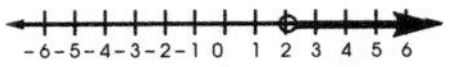

2. $^-15m \leq ^-75$ $m \geq 5$

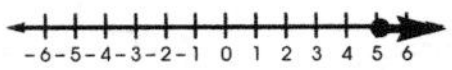

3. $^-1 > \frac{b}{3}$ $b < ^-3$

4. $1.9x \leq ^-7.6$ $x \leq ^-4$

5. $\frac{3}{2}y < 6$ $y < 4$

6. $^-26m \geq 13$ $m \leq \frac{^-1}{2}$

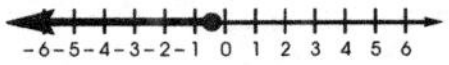

7. $^-4 \geq \frac{2}{3}x$ $x \leq ^-6$

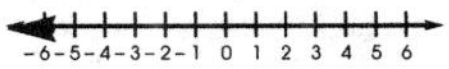

8. $^-2c < 2$ $c > ^-1$

9. $^-3a \leq ^-9$ $a \geq 3$

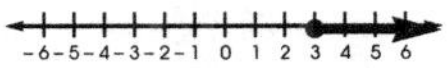

10. $-\frac{3}{4}x \geq ^-3$ $x \leq 4$

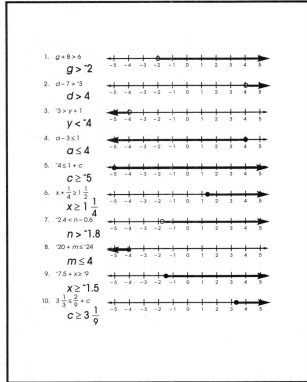

86

Answer Key

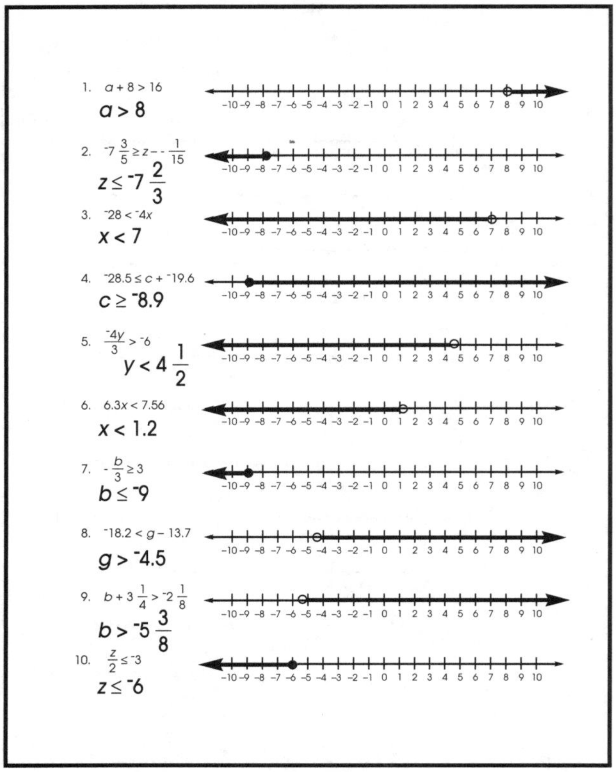

1. $a + 8 > 16$
   $a > 8$
2. $^-7\frac{3}{5} \ge z - ^-\frac{1}{15}$
   $z \le {}^-7\frac{2}{3}$
3. $^-28 < {}^-4x$
   $x < 7$
4. $^-28.5 \le c + {}^-19.6$
   $c \ge {}^-8.9$
5. $\frac{^-4y}{3} > {}^-6$
   $y < 4\frac{1}{2}$
6. $6.3x < 7.56$
   $x < 1.2$
7. $-\frac{b}{3} \ge 3$
   $b \le {}^-9$
8. $^-18.2 < g - 13.7$
   $g > {}^-4.5$
9. $b + 3\frac{1}{4} > {}^-2\frac{1}{8}$
   $b > {}^-5\frac{3}{8}$
10. $\frac{z}{2} \le {}^-3$
    $z \le {}^-6$

87

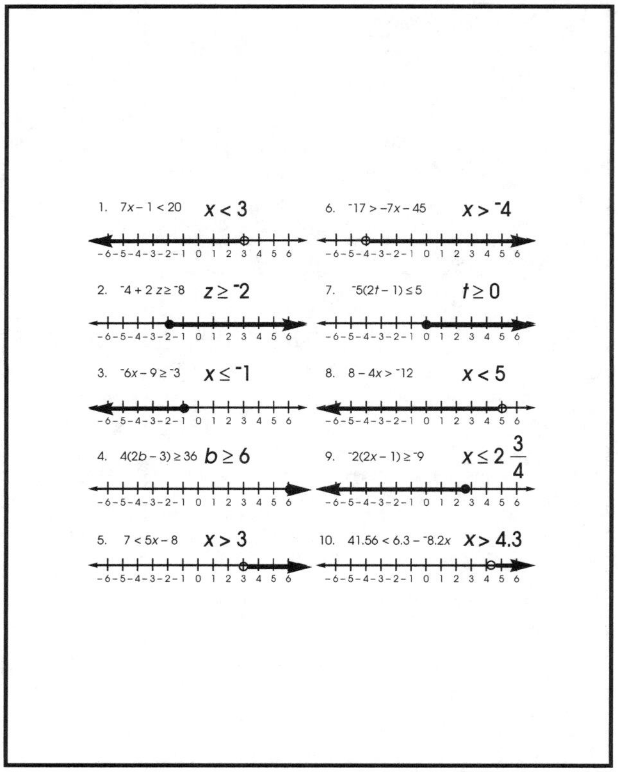

1. $7x - 1 < 20$ $x < 3$
2. $^-4 + 2z \ge {}^-8$ $z \ge {}^-2$
3. $^-6x - 9 \ge {}^-3$ $x \le {}^-1$
4. $4(2b - 3) \ge 36$ $b \ge 6$
5. $7 < 5x - 8$ $x > 3$
6. $^-17 > -7x - 45$ $x > {}^-4$
7. $^-5(2t - 1) \le 5$ $t \ge 0$
8. $8 - 4x > {}^-12$ $x < 5$
9. $^-2(2x - 1) \ge {}^-9$ $x \le 2\frac{3}{4}$
10. $41.56 < 6.3 - {}^-8.2x$ $x > 4.3$

88

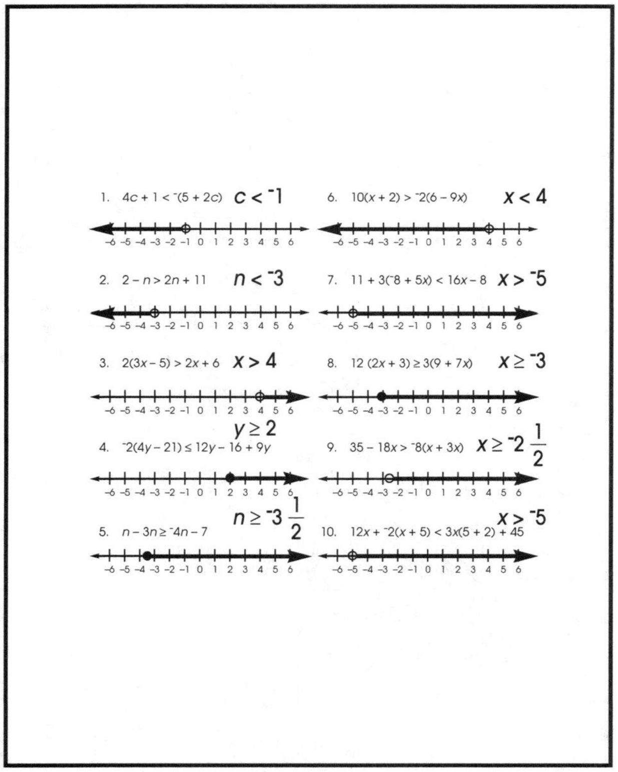

1. $4c + 1 < {}^-(5 + 2c)$ $c < {}^-1$
2. $2 - n > 2n + 11$ $n < {}^-3$
3. $2(3x - 5) > 2x + 6$ $x > 4$
4. $^-2(4y - 21) \le 12y - 16 + 9y$ $y \ge 2$
5. $n - 3n \ge {}^-4n - 7$ $n \ge {}^-3\frac{1}{2}$
6. $10(x + 2) > {}^-2(6 - 9x)$ $x < 4$
7. $11 + 3({}^-8 + 5x) < 16x - 8$ $x > {}^-5$
8. $12(2x + 3) \ge 3(9 + 7x)$ $x \ge {}^-3$
9. $35 - 18x > {}^-8(x + 3x)$ $x \ge {}^-2\frac{1}{2}$
10. $12x + {}^-2(x + 5) < 3x(5 + 2) + 45$ $x > {}^-5$

89

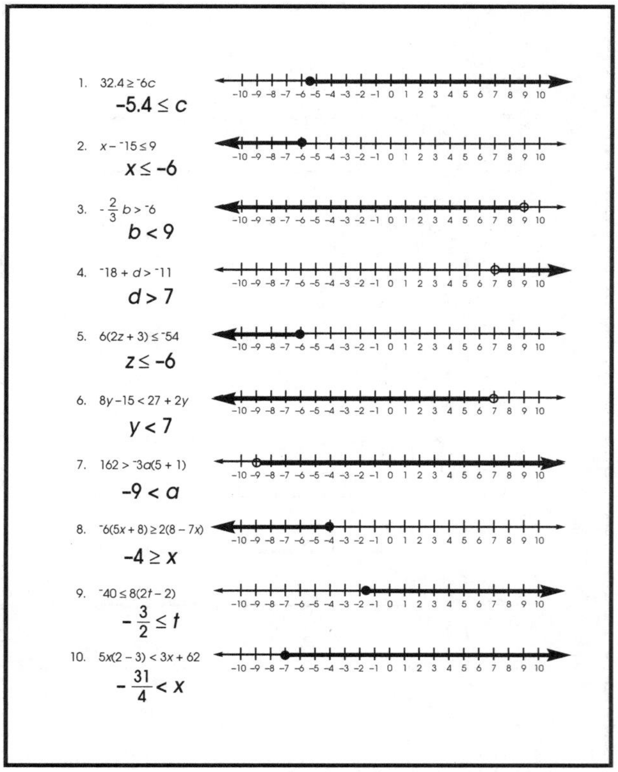

1. $32.4 \ge {}^-6c$
   $-5.4 \le c$
2. $x - {}^-15 \le 9$
   $x \le -6$
3. $-\frac{2}{3}b > {}^-6$
   $b < 9$
4. $^-18 + d > {}^-11$
   $d > 7$
5. $6(2z + 3) \le {}^-54$
   $z \le -6$
6. $8y - 15 < 27 + 2y$
   $y < 7$
7. $162 > {}^-3a(5 + 1)$
   $-9 < a$
8. $^-6(5x + 8) \ge 2(8 - 7x)$
   $-4 \ge x$
9. $^-40 \le 8(2t - 2)$
   $-\frac{3}{2} \le t$
10. $5x(2 - 3) < 3x + 62$
    $-\frac{31}{4} < x$

90

© Carson-Dellosa • CD-704384

Answer Key

1. $9x - 8 + x < 16 + 4x$

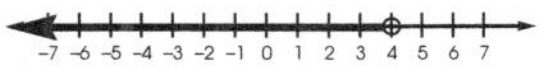

2. $15y \geq {}^-45$

3. $69 > c + 71$

4. $17 + 11n - 13 \leq 4(n + 1) + 2n$

5. $8(2 + x) > 3(x - 3)$

6. ${}^-4(3x + 2) \geq 40$

7. $\frac{5}{3} < \frac{2}{3}x - 1$

8. $3n - 4(2n - 5) + n + 4 \geq 0$

9. $18c + 11 - 26c < -3c(5 + 1) - 59$

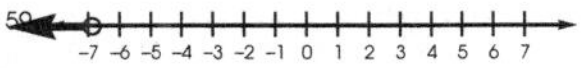

10. $8a - 2(2a + 5) \leq 2a(9 + 1) + 54$

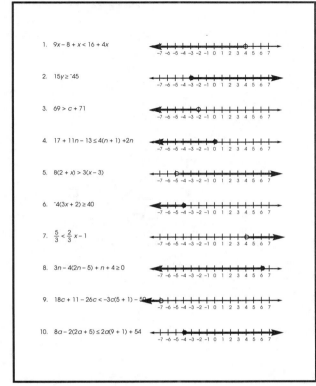

91

1. $a + {}^-7 = 8$	15	8. ${}^-27 = c + 27$	$^-54$
2. $y + 76 = {}^-93$	$^-169$	9. $-\frac{5}{8} + x = -\frac{5}{8}$	0
3. $4 + b = {}^-14$	$^-18$	10. $y + {}^-6.2 = 8.1$	14.3
4. ${}^-33 = z + 16$	$^-49$	11. $38 = x + {}^-19$	57
5. ${}^-12 + x = 21$	33	12. $a + {}^-2\frac{5}{9} = {}^-10\frac{5}{18}$	$^-7\frac{13}{18}$
6. $2.4 = m + 3.7$	$^-1.3$	13. ${}^-1{,}129 + b = 3{,}331$	4460
7. ${}^-1\frac{1}{2} + n = {}^-1\frac{5}{8}$	$-\frac{1}{8}$	14. ${}^-3.5 = 7\frac{1}{2} + x$	$^-11$

92

1. $k - 36 = 37$	73	8. ${}^-17 = q - 3$	$^-14$
2. ${}^-22 = y - 8$	$^-14$	9. $p - \frac{3}{5} = \frac{3}{5}$	$\frac{6}{5}$
3. $x - {}^-7 = {}^-19$	$^-26$	10. $5.62 = m - 6$	11.62
4. $30 = b - {}^-2$	28	11. $x - {}^-36.5 = {}^-2.563$	$^-39.063$
5. $a - 18 = {}^-32$	$^-14$	12. ${}^-1{,}132 = b - 6{,}339$	5207
6. ${}^-1.7 = b - 9.3$	7.6	13. $7\frac{3}{4} = a - 16\frac{3}{16}$	$23\frac{15}{16}$
7. ${}^-4\frac{1}{3} = q - 3\frac{1}{3}$	$^-1$	14. $z - {}^-5.75 = {}^-8\frac{1}{4}$	$^-14$

93

1. $x + {}^-3 = -18$	$^-15$	11. ${}^-8.6 = m + 11.12$	$^-19.72$
2. $c - 11 = 43$	54	12. $a - -\frac{1}{5} = \frac{3}{20}$	$-\frac{1}{20}$
3. $12 + y = 32$	20	13. $-\frac{3}{4} + z = \frac{7}{18}$	$\frac{41}{36}$
4. ${}^-26 = d - 7$	$^-19$	14. $b - 17.8 = {}^-36$	$^-18.2$
5. ${}^-62 = a + 16$	$^-78$	15. $-\frac{13}{24} = -\frac{5}{16} + c$	$-\frac{11}{48}$
6. $q - {}^-83 = 121$	38	16. $102.8 = g - {}^-66.09$	36.71
7. $t + {}^-101 = 263$	364	17. $f + \frac{3}{5} = \frac{3}{4}$	$\frac{3}{20}$
8. $w - 454 = {}^-832$	$^-378$	18. $b - \frac{5}{6} = -\frac{7}{8}$	$-\frac{1}{24}$
9. ${}^-332 = {}^-129 + s$	$^-203$	19. $21.21 + p = {}^-101.6$	$^-122.81$
10. $665 = k - {}^-327$	338	20. ${}^-762.46 = h - 32.061$	$^-730.399$

94

Answer Key

#	Problem	Answer
1.	$^{-}6a = ^{-}66$	11
2.	$^{-}180 = 12b$	$^{-}15$
3.	$^{-}13n = 13$	$^{-}1$
4.	$42 = ^{-}14p$	$^{-}3$
5.	$1\frac{1}{2} = 3x$	$\frac{1}{2}$
6.	$^{-}5.6 = ^{-}0.8x$	7
7.	$8 = ^{-}32b$	$-\frac{1}{4}$
8.	$9a = ^{-}3$	$-\frac{1}{3}$
9.	$0.25y = 1.5$	6
10.	$^{-}0.0006 = 0.02x$	$^{-}0.03$
11.	$^{-}11x = 275$	$^{-}25$
12.	$45\frac{1}{2} = ^{-}14c$	$^{-}3\frac{1}{4}$
13.	$61.44 = 12z$	5.12
14.	$^{-}21y = ^{-}756$	36

95

#	Problem	Answer
1.	$^{-}18 = \frac{a}{6}$	$^{-}108$
2.	$\frac{x}{6} = ^{-}6$	$^{-}36$
3.	$\frac{y}{^{-}2} = 231$	$^{-}462$
4.	$\frac{1}{5}b = ^{-}8$	$^{-}40$
5.	$\frac{m}{0.6} = 0.3$	0.18
6.	$35 = \frac{x}{^{-}7}$	$^{-}245$
7.	$0.12 = \frac{y}{0.12}$	0.0144
8.	$3 = -\frac{1}{8}a$	$^{-}24$
9.	$\frac{w}{^{-}2} = 0.04$	$^{-}0.08$
10.	$\frac{u}{^{-}4} = ^{-}14$	56
11.	$\frac{x}{^{-}5.1} = ^{-}16$	81.6
12.	$-28 = \frac{a}{13}$	$^{-}364$
13.	$\frac{1}{18}c = ^{-}31$	$^{-}558$
14.	$\frac{b}{^{-}0.29} = 5.5$	$^{-}1.595$

96

#	Problem	Answer
1.	$^{-}2p = ^{-}38$	19
2.	$\frac{b}{8} = ^{-}24$	$^{-}192$
3.	$^{-}85 = 17r$	$^{-}5$
4.	$^{-}32 = \frac{c}{^{-}22}$	704
5.	$^{-}13a = 52$	$^{-}4$
6.	$\frac{1}{47}d = ^{-}26$	$^{-}1222$
7.	$^{-}12f = ^{-}180$	15
8.	$\frac{1}{0.16}x = 0.7$	0.112
9.	$^{-}77.4 = 9a$	$^{-}8.6$
10.	$-\frac{1}{6}q = ^{-}11$	66
11.	$16 = \frac{n}{^{-}21}$	$^{-}336$
12.	$0.7h = ^{-}0.112$	$^{-}0.16$
13.	$^{-}80 = \frac{p}{15}$	$^{-}1200$
14.	$792 = ^{-}33y$	$^{-}24$
15.	$^{-}5.2 = \frac{m}{30.1}$	$^{-}156.52$
16.	$^{-}11.2x = ^{-}60.48$	5.4
17.	$\frac{1}{^{-}26}r = ^{-}66$	1716
18.	$315 = 21s$	15
19.	$\frac{z}{0.06} = ^{-}7.98$	$^{-}0.4788$
20.	$^{-}14g = ^{-}406$	29

97

#	Problem	Answer
1.	$13 + ^{-}3p = ^{-}2$	5
2.	$\frac{^{-}5a}{2} = 75$	$^{-}30$
3.	$6x - 4 = ^{-}10$	$^{-}1$
4.	$9 = 2y + 9$	0
5.	$^{-}10 + \frac{a}{4} = 9$	76
6.	$17 = 5 - x$	$^{-}12$
7.	$^{-}7r - 8 = ^{-}14$	$\frac{6}{7}$
8.	$\frac{4y}{3} = 8$	6
9.	$16 + \frac{x}{3} = ^{-}10$	$^{-}78$
10.	$\frac{^{-}4z}{5} = -12$	15
11.	$^{-}22 = 3s - ^{-}8$	$^{-}10$
12.	$-\frac{a}{6} - ^{-}31 = 64$	$^{-}198$

98

© Carson-Dellosa • CD-704384

Answer Key

1. $x = 8$ $\frac{x}{4} + 5 = 7$	2. $x = 15$ $2x - 20 = 10$	3. $x = {}^-8$ ${}^-3x - 12 = 12$	4. $x = {}^-1$ $-x - 6 = -5$	5. $x = 6$ $2 = 2x - 10$
6. $x = 14$ $3x - 7 = 35$	7. $x = {}^-4$ $2 + 5x = {}^-18$	8. $x = {}^-2$ $4x + 5 = {}^-3$	9. $x = 5$ ${}^-11x + 10 = {}^-45$	10. $x = 7$ $5x - 6 = 29$
11. $x = {}^-5$ ${}^-4 = \frac{4x}{5}$	12. $x = {}^-3$ ${}^-2x + 7 = 13$	13. $x = 4$ $\frac{5x}{4} + 2 = 7$	14. $x = 11$ ${}^-64 = {}^-5x - 9$	15. $x = 13$ $2x + 10 = 36$
16. $x = 1$ $8x - 9 = {}^-1$	17. $x = 3$ $12 = 3x + 3$	18 $x = 10$ ${}^-4 = \frac{2x}{{}^-5}$	19. $x = 12$ $9 + 4x = 57$	20. $x = {}^-6$ $\frac{x}{{}^-2} + 2 = 5$
21. $x = 2$ $6 = \frac{x}{2} + 5$	22. $x = 9$ $2x - 10 = 8$	23 $x = 16$ $8 = \frac{x}{4} + 4$	24. $x = {}^-7$ $5x - 15 = {}^-50$	25. $x = 0$ $3x - 9 = {}^-9$

The Magic Sum is 20.

99

1. $4x - 7 = 2x + 15$ **11**
2. ${}^-4 = {}^-4(f - 7)$ **8**
3. $5x - 17 = 4x + 36$ **53**
4. $3(k + 5) = {}^-18$ **${}^-11$**
5. $y + 3 = 7y - 21$ **4**
6. ${}^-3(m - 2) = 12$ **${}^-2$**
7. $18 + 4p = 6p + 12$ **3**
8. ${}^-8\,(\frac{a}{8} - 2) = 26$ **${}^-10$**
9. ${}^-3k + 10 = k + 2$ **2**
10. $22 = 2\,(b + 3)$ **8**
11. $6a + 9 = {}^-4a + 29$ **2**
12. ${}^-22 = 11(2c + 8)$ **${}^-5$**
13. $10p - 14 = 9p + 17$ **31**
14. ${}^-45 = 5(\frac{2a}{5} + {}^-3)$ **${}^-15$**
15. $16z - 15 = 13z$ **5**
16. $36 + 19b = 24b + 6$ **6**
17. $144 = {}^-16\,(3 + 3d)$ **${}^-4$**
18. $11h - 14 = 7 + 14h$ **${}^-7$**
19. ${}^-3\,(\frac{2j}{3} - 6) = 32$ **${}^-7$**
20. ${}^-43 - 3z = 2 - 6z$ **15**

100

Solve these equations.

1. ${}^-116 = -a$ — $a = 116$
2. $6m - 2 = m + 13$ — $m = 3$
3. $x + 2 = {}^-61$ — $x = {}^-63$
4. ${}^-30 = {}^-6 - y$ — $y = 12$
5. ${}^-5t + 16 = {}^-59$ — $t = 15$
6. $4a - 9 = 6a + 7$ — $a = {}^-8$
7. $\frac{{}^-3b}{8} = {}^-36$ — $b = 96$
8. ${}^-40 = 10\,(4 + s)$ — $s = {}^-8$
9. $28 - \frac{k}{3} = 16$ — $k = 36$
10. ${}^-9r = 20 + r$ — $r = {}^-2$
11. $114 = 11c - {}^-26$ — $c = 8$
12. ${}^-38 = 17 - 5z$ — $z = 11$
13. ${}^-5(2x - 5) = {}^-35$ — $x = 6$
14. $20c + 5 = 5c + 65$ — $c = 4$
15. $\frac{{}^-d}{5} - 21 = {}^-62$ — $d = 205$
16. $\frac{{}^-15c}{{}^-4} = {}^-30$ — $c = {}^-8$
17. $384 = 12\,({}^-8 + 5t)$ — $t = 8$
18. $3n + 7 = 7n - 13$ — $n = 5$
19. ${}^-8 - \frac{y}{3} = 22$ — $y = {}^-90$
20. ${}^-5t - 30 = {}^-60$ — $t = 6$

HINT: The sum of the solutions equals the number of steps in the Statue of Liberty — 354

101

1. Three-fifths of a number decreased by one is twenty-three. What is the number?
   40
2. Seven more than six times a number is negative forty-seven. What is the number?
   ${}^-9$
3. Nine less than twice a number is thirty-one. What is the number?
   20
4. Three times the sum of a number and five times the number is thirty-six. What is the number?
   2
5. The quotient of a number and four decreased by ten is two. What is the number?
   48
6. Carol is sixty-six inches tall. This is twenty inches less than two times Mindy's height. How tall is Mindy?
   43 inches
7. In February, Paul's electric bill was three dollars more than one-half his gas bill. If the electric bill was ninety-two dollars, what was the gas bill?
   $178

102

1. One of two numbers is five more than the other. The sum of the numbers is 17. Find the numbers.
   6 and 11
2. The sum of two numbers is twenty-four. The larger number is three times the smaller number. Find the numbers.
   6 and 18
3. One of two numbers is two-thirds the other number. The sum of the numbers is 45. Find the numbers.
   27 and 18
4. The difference of two numbers is 19. The larger number is 3 more than twice the smaller. Find the numbers.
   16 and 35
5. 320 tickets were sold to the school play. There were three times as many student tickets sold as adult tickets. Find the number of each.
   80 adults 240 students
6. The first number is eight more than the second number. Three times the second number plus twice the first number is equal to 26. Find the numbers.
   10 and 2
7. Dan has five times as many $1 bills as $5 bills. He has a total of 48 bills. How many of each does he have?
   8 – $5 bills 40 – $1 bills

103

© Carson-Dellosa • CD-704384